A Bouquet for the G

A Bouquet for the Gardener

Martin Gardner Remembered

The Lewis Carroll Society of North America

Photographs of Martin Gardner courtesy of, and ©2011, Jim Gardner
Page viii: photo by George Cserna
Pages 14, 105, 188: photos by David B. Eisendrath, Jr.

Designed and typeset with Adobe InDesign by Andrew Ogus
Edited by Mark Burstein

ISBN 978-0-930326-17-3

The Lewis Carroll Society of North America
www.LewisCarroll.org

Cover designed by Andrew Ogus
Gardner portrait by Mahendra Singh
Rose images by Wernher Krutein / Photovault.com
Ambigrams by Scott Kim

Contents

. . . makes hobby-hodge happy in his hole.[2]

[2] I have heard this word used by Martin Halpin, an old gardener from the Glens of Antrim . . .

—James Joyce, *Finnegans Wake*, 1939

Foreword

JIM GARDNER

Over the course of Dad's career, he contributed across a variety of interesting areas: math, science, philosophy, magic, and literature. Dad's friend Douglas Hofstadter, the cognitive and computer scientist, once commented on how underwhelming our home was with regard to the display of products related to math and puzzles. A similar observation could be made with regard to Alice. If you walked into our house, there were no reproductions of Tenniel's illustrations or Lewis Carroll–related memorabilia prominently on display. In the latter years of Dad's life there were more obvious graphics hanging on the wall of his apartment, such as an enlarged, framed cover from the British editions of *More Annotated Alice*, and a photo of Dad and the Alice sculpture in Central Park, New York, that was part of a set taken for the back cover of the original edition. But for the most part, visitors to our home who were unfamiliar with his work would not have deduced he was an admirer of Lewis Carroll.

In what manner, then, did Lewis Carroll reside in the life of Martin Gardner? Like much of his other work, it was prominently in his mind

Dr. James Emmett (Jim) Gardner is a Professor in the Special Education Program, Department of Educational Psychology at The Jeannine Rainbolt College of Education of The University of Oklahoma in Norman, and the editor of the scholarly Journal of Special Education Technology. *Martin's last years were spent in Norman in order to be near him.*

and his approach to things. I am not a Carroll scholar (sorry, everyone, but I grew up reading L. Frank Baum's Oz series), thus I can't speak with much authority or insight cross-referencing Dad's and Carroll's writings. However, I can speak to the fact that there clearly were few things more prominent regarding Dad's alignment to fictional literature than his love of Alice and Carroll's other work. Dad clearly had a whimsical side to him, and I'm certain that satire and imaginative elements of *Alice*, *The Snark* (Dad's other annotated Carroll work), and "Jabberwocky" gave him great pleasure.

Those who knew Dad well knew that he was nearly impeccable in regard to correspondence through the mail. Those who were even closer knew that at times he loathed getting fan mail from strangers, simply and almost entirely because it meant composing yet another letter of reply. However, from a more pragmatic perspective, Dad actually had great admiration for his Alice fan base. For example, I remember a conversation we had, subsequent to the timely 2010 release of Tim Burton's movie and *The Annotated Alice*'s "Easter egg" appearance in season six of *Lost*. He remarked, "You know (gesturing with a finger towards the file cabinet that was next to his desk), I have a file full of new annotations that readers have sent me since the Definitive edition. I probably have enough for a fourth edition or more!" Dad subsequently described an example of a communiqué he had received from his fellow magician, Alice admirer, and skeptical friend, Teller.

When Alice meets the Caterpillar, the Caterpillar remarks, "Who are YOU?" Apparently, Teller had been reading Charles Mackay's *Extraordinary Popular Delusions and the Madness of Crowds*, and came across a section describing that a popular gag line in London around the time of Alice's composition, which was to quickly utter the remark "*Who* are *you*?!" I could clearly see the sparkle in Dad's eyes as he recounted this new discovery, which had yielded yet another tidbit to deciphering the underlying, subtle humor of *Alice* as it might have been read by Carroll's contemporaries in the 1860s.

Growing up on 10 Euclid Ave, in Hastings-on-Hudson, the morning paper was *The New York Times* and the afternoon paper was the *Herald Tribune*. Of course, the morning *Times* was *the* delivery of the news to the house, but I never really understood why we received the *Tribune*. When I was in the seventh or eighth grade, for a few months I was our street's local paper boy. During this phase I insisted on—to use a Las Vegas term—"comping" my parents' newspaper. When I informed my parents that I was quitting my route and they would have to begin paying for the *Tribune,* Dad remarked that the *only* reason he subscribed to the *Tribune* was because it published a daily cryptogram puzzle that he liked to decipher! (Many years later I learned from Dad that he had contacted the *Herald Tribune*—likely by letter—to inquire whether they would be willing to simply deliver him the page containing the daily cryptogram. I have a vague recollection that one of the editors at the *Herald* recognized who Dad was, and used it as an opportunity to deliver the bad news as part of a "nice chat." But this anecdote is more lore about Martin Gardner, and I can't attest to the correspondence.) This is my way of contextualizing why I specifically chose the word "decipher" in a preceding sentence, because my sense is that for a major segment in Dad's life, one of his principal and most relished recreations was deciphering the explicit—and concealed—puzzles and hidden meaning scattered across Carroll's writing. Indeed, Dad's books *The Snark Puzzle Book* and *The Universe in a Handkerchief* (the Preface of which includes Dad's brief testament to how he realized there was much more to Carroll's writing than merely a child's adventure and some entertaining nonsensical poems) demonstrate how he had the natural gift of intuition when it came to annotating or repurposing the conundrum side of Carroll's work.

If one were to ask Dad what *he* considered his greatest writing accomplishment, I'm honestly not sure what area or specific publication he would name. I suppose this may be an interesting topic for future scholars and students in seminars, or friends and colleagues at cocktail parties or gatherings. But I do know that in the months leading up to his death, he wavered between expressing perplexed astonishment and tantalizing pride regarding the enduring interest in and sales of *The Annotated Alice.* It leads me wonder whether there is a form of perpetual motion in the universe, after all.

Introduction

MARK BURSTEIN, EDITOR

Traditionally, one inserts the word "arguably" or the like when making an aesthetic judgment, but I cannot imagine anyone disagreeing with the statement, "The most important edition of Carroll's masterworks after their initial issues of 1865/66 and 1871/72 was Martin Gardner's *The Annotated Alice* of 1960." Although by no means the first book to include critical apparatus such as footnotes, endnotes, and marginalia, his pioneering format, extensive research, keen judgment, and generous commentary provided readers with contexts, comparisons, alternatives, and explanations, entertaining and enlightening readers on a scale never before available, and incidentally paving the way for a host of other annotated editions of canonical works. (One must wonder how much this influenced Vladimir Nabokov, whose *Pale Fire*, a volume in which the commentary, most drolly, far overshadows the text, was published two years later.)

Mark Burstein is the president of the Lewis Carroll Society of North America, a body he has previously served as "Warden" and co-founder, with his father, Sandor, of its West Coast Chapter (1979 to 1987); editor of its magazine, Knight Letter *(1994 to 2006), currently its production editor; and vice president (1999 to 2006). A book editor by trade, his collaborations include* Lewis Carroll's La Guida di Bragia, *Linda Sunshine's* All Things Alice, Pictures and Conversations: Lewis Carroll in the Comics, *and* Alice Illustrated, *coming from Dover this fall.*

The very incarnation of the word polymath (pun unavoidable) and a walking contradiction, Gardner, the twentieth century's most famous explicator of mathematics for the masses, never took a math class beyond high school, never used email, and browsed the Internet only rarely. A famous skeptic and debunker of pseudoscience, he nevertheless declared himself to be a believer in God, based on faith alone. Possessing a mind that reached the outer shores of our universe, he was born, and died, in Oklahoma. One of the most famous authors and popularizers of science and math in an age that worships celebrity, he was notoriously reluctant to appear in public or to exploit his fame.

It could not be more appropriate that the epigraph to this book comes from a singular chapter of *Finnegans Wake*—a book about which Gardner wrote extensively and in which Carroll often appears—that not only features footnotes and marginalia, but whose very foundation is a geometrical problem to be solved! Although but a recent college grad at the time the *Wake* was published, his name seems to magically appear on the page; naturally it was in an annotation. The quote further fits on many levels beyond the obvious rabbit "hole," including his hodgepodge of hobbies; "hodgee," which the *OED* defines as a "teacher"; and the mention of "Antrim," a county in Northern Ireland, which happens to be an anagram of "Martin," and anagrams were a favorite of Gardner's. Instances of bibliomancy and astonishing coincidence abound in the text of the *Wake*, but I can imagine Martin's eyebrows beginning to arch here, so I will leave that point for another time.

Like most everyone else, I have always viewed Martin Gardner as a somewhat mythical figure. I devoured his *Scientific American* "Mathematical Games" column each month (I was six when it started, so I doubt I understood much of it then, but it was a delight to look at), and I'm sure my father brought home a copy of *The Annotated Alice* the moment it was in the bookstores (I was ten). It never crossed my mind to write to him.

The first time I heard from Martin was in the fall of '97, after I had been editing the *Knight Letter* for a few years. His letters were always unfailingly polite, and he habitually used a typewriter, hand-underlining appropriate words, and making corrections in his own hand, in ink. Several of his letters were printed in our magazine, and we continued a sporadic correspondence over the years, until the day in May of '05 when I received a manila envelope in the mail. The enclosed letter began, "Dear Mark, is the enclosed document suitable for *Knight Letter*?" Entitled "A Supplement to *Annotated Alice*," it contained fourteen typewritten pages of further annotations and revisions, proving, among other things, that he had never stopped thinking about Carroll. "*Suitable*"? Yes, I think so. These appeared as the featured article in issue 75 of the *Knight Letter* (Summer, 2005), and he soon sent another batch, which appeared in the next issue (Spring, 2006). These sets of annotations, which are reproduced in this present volume, proved to be his last.

I mentioned in a letter to him around this time that he was no small part of my decision to name my son Martin, to which he modestly replied that he was "honored and a bit overwhelmed to learn that I played a role in your son's name." I feel likewise honored, and more than a bit overwhelmed, to have had a hand in this volume.

A founding member, with his beloved wife, Charlotte, of our Society, Martin attended our now legendary first meeting, at the Nassau Inn in Princeton, in January of 1974, and several in the years thereafter. Our debt to him is enormous, and it is our hope that this book may in some small way start to offset that debt, at least to his family, friends, colleagues, and admirers. He himself would have been embarrassed by the fuss.

As always with our Society, this book reflects the work of many. The initial brainstorming for this volume came from Dr. Francine Abeles, August and Clare Imholtz, Andrew Sellon, and myself. We decided it was to be a portmanteau, with both an academic festschrift, a series of essays done in his honor, and a section of tributes and reminiscences by those who knew, or were greatly influenced by, him. Our contributors include Carrollians, mathematicians, magicians, authors, academics, and a wide assortment of others. Andrew Ogus graciously contributed his design talents.

Ad majorem Hortulani gloriam!

IN MEMORIAM

Martin Gardner & the Grown-Ups

WILL BROOKER

I first encountered Alice, appropriately, in reverse: my mother gave me her copy of *Through the Looking-Glass* when I was eight years old, with the words "your turn now" written inside the cover. Carroll, who wrote in his "Easter Greeting" of a mother's gentle hand drawing the curtains, and her voice summoning the child to rise, would surely have approved of the gift; and it might also have pleased him that, with Looking-Glass logic, I approached *Alice's Adventures in Wonderland*, some months later, as the sequel to a story I'd come to love.

As an eight-year old, I was inquisitive, logical, imaginative but practical, and so I preferred the structured, cooler, slightly more rational nonsense of *Looking-Glass*—with its railways, rules, grids, and codes—to the looser play of *Wonderland*. I remember the world as a series of mysteries whose answers were just out of reach; mysteries I knew would become clear to me when I was a grown-up, without being sure exactly when that would be, or being patient enough to wait. So I puzzled and worried at the boundaries of my own knowledge, figuring some things out and making up others (I

Will Brooker is Reader in (Professor of) Film and Television at Kingston University, London. He is the author of several books on popular culture and audience, including Batman Unmasked; Using the Force: Creativity, Community, and Star Wars Fans; The Blade Runner Experience; BFI Film Classics: Star Wars; *and* Alice's Adventures: Lewis Carroll in Popular Culture.

remember my parents' patient correction when I told them I'd realized the "guerrilla fighters" from the TV news were humans who hadn't evolved fully from apes). The *Alice* books, and *Looking-Glass* in particular, were perfect analogies for my way of seeing the world, the larger world of adults, because of course Alice is a seven-year-old (or seven-and-a-half in *Looking-Glass*) making her way not through a land of other children, but the society of adults. So she encounters poems she doesn't understand, and asks for translations only after stubbornly struggling to make sense of them herself ("You see," Carroll adds wryly, "she didn't like to confess, even to herself, that she couldn't make it out at all"). Alice is a proud, curious, and clever little girl, but she still asks Humpty to explain the meaning of words.

One of those words is "impenetrability," a word Martin Gardner must have thought fitting in association with *Alice*. With some regret, in his Introduction to *The Annotated Alice*, he suggests that Carroll's nonsense will seem random and pointless to the average modern youngster: " . . . the time is past when a child under fifteen, even in England, can read *Alice* with the same delight as gained from, say, *The Wind in the Willows* or *The Wizard of Oz*." My mother gave me *The Annotated Alice* when I was ten and, as stubborn and proud as Alice herself, I took that as a challenge.

Already, Carroll's books had tested and pushed me, as well as delighted me. They had been, in some ways, like the best English lesson on Earth (with a bit of mathematics thrown in—extra); they had made me think about the meanings of words, their wriggling, lizard-like nature, and the games they could play. But Gardner was right: there were some things that even a clever child of the 1970s couldn't hope to decipher from two books published in 1865 and 1871. There were, in fact, many things. There were other levels, other dimensions, other worlds—darker worlds, more complex, more difficult, and more rewarding worlds—hidden beneath the fairy tales. I had always suspected as much, just as Alice did about "Jabberwocky" ("Somehow it seems to fill my head with ideas—only I don't exactly know what they are!"), but Gardner revealed them, and explained them. He transformed the *Alice* stories into an encyclopedia. If the *Alice* adventures were like a wonderful schoolroom with Carroll as the conjuror-teacher, *The Annotated Alice* was a university in a book, and Gardner a wise and witty professor.

As his Introduction plainly states, Gardner's notes are addressed not to children but to adult readers and in many ways, this was their appeal to me as a ten-year-old. Gardner did not talk down to me; he assumed I had certain knowledge, and made me run after him, rather than slowing for me to catch up. When his references baffled me I either had to guess, or struggle for the meaning, or ask someone, or find a dictionary, or—as instant explanations were hard to come by, so many years before the Internet—assume I'd find out later, when I was older. But even when I barely understood his explanations, they let me glimpse that world of adult knowledge and understanding. They let me into previously hidden rooms where I walked around freely, following his tour and picking up what I could. Gardner guided me into places and areas of knowledge I'd only guessed at. After reading *The Annotated Alice*, I knew about base 18 and base 21. I knew about caucuses. I knew about G. K. Chesterton and O. Henry. I knew that the name "passion flower" referred to the Passion of Christ. I knew what an antimacassar was. I knew about *Finnegans Wake*. I was still only ten years old.

Thanks to Martin Gardner and Lewis Carroll—and thanks to my parents, not least for buying me those books—I think I knew more at that age than I do now; my head was so full of shiny, slippery ideas that most could only escape, like fish from a kettle, as the years passed. Some of them, though, I grasped more tightly and studied further. Thanks to Gardner, I studied Joyce and read Chesterton, and enjoyed Kerouac, and learned chess. More fundamentally, it was partly through Gardner I learned the joy of close analysis and research, in making links and connecting sources, that shaped my later career as an academic. My book *Alice's Adventures: Lewis Carroll in Popular Culture* is, I think, infused with the pleasure in pedantry that I first encountered in *The Annotated Alice*. I still know as much, or as little about base 18 as I knew at age ten, but thanks to Martin Gardner, I know it is there, in a little room, waiting for me to sit down one day and study it. With the confidence of Humpty Dumpty but the kindness of the White Knight, Martin Gardner showed me what it could be like to be a grown-up. He was the best type of grown-up.

My Pen Pal, Martin Gardner

ANGELICA CARPENTER

Like many other researchers and fans, I am proud to say that Martin Gardner was my pen pal. We corresponded for ten years about Alice and Oz.

The relationship began badly in 1999 when I panned his novel *Visitors from Oz* in the *Lewis Carroll Review*. Martin wrote a rebuttal in the next issue. I was amazed that he had responded. I felt terrible to have upset a person whose work I knew and admired.

Martin was a charter member of the International Wizard of Oz Club,[1] and I joined in the 1970s, after learning of the Club's existence through *The Annotated Wizard of Oz*. I knew of Martin's role in getting that book published, of convincing his publisher to hire Michael Patrick Hearn to write a companion book to Martin's wonderful *Annotated Alice*. I loved the *Alice* book, too, and the genre that Martin had re-invented.

I was sure that I had ruined any chance of ever connecting with him, but Martin didn't hold a grudge. Soon he praised me in another publication for some Ozzy effort. I wrote to thank him, and we got off to a second, much

Angelica Carpenter is curator of the Arne Nixon Center for the Study of Children's Literature at California State University, Fresno. The author or co-author of four biographies for young people, Frances Hodgson Burnett, L. Frank Baum, Robert Louis Stevenson, *and* Lewis Carroll, *she also edited a scholarly anthology,* In the Garden: Essays in Honor of Frances Hodgson Burnett. *She is past president of the International Wizard of Oz Club and a board member of the Lewis Carroll Society of North America.*

better start. I wish that I had saved his letters. I know that I invited him to speak at the Oz Centennial Convention in 2000, and that he declined, not liking to speak in such settings (though he did attend some early Oz gatherings).

In 2005, as president of the Oz Club, I asked Martin to write his reminiscences for a 50th-anniversary issue of our magazine, *The Baum Bugle*. We talked then, on the phone. He didn't use a computer, he explained, but he soon mailed a wonderful article and it's the one letter that I did save.

"Forgive my terrible typing," he wrote. "I could retype, but then I would make even new errors. I enjoyed writing it. It brought back a flood of memories."

The Bugle was behind schedule then, so he called a couple of times to see when his article would be published. "After all, I am 92," he explained. He was 93 when it finally appeared in 2007.

I watched a television interview with him made that same year, which was shown at an Oz convention. It was an informal session, recorded on home video by Oz Club member Kevin Thomas. Martin was simply terrific—relaxed and articulate as he called up precise memories from half a century of Oz. Not so surprising, I guess, from an author who had just published a major new book (*The Annotated Hunting of the Snark*), but I don't know anyone else, of any age, who could have done it so well.

He found pleasure in many things, even living in small quarters, in an assisted living facility. He had what he needed, he told me, and he liked being near his son. His mind and his imagination were unrestricted and his sense of wonder was intact. "I never cease to be amazed," he wrote, "at the subtleties still being discovered in the *Alice* books."

It would be impossible to know how many readers have found their ways to Oz and Wonderland on paths revealed by Martin Gardner.

[1] Founded as the "Wizard of Oz Fan Club" in 1957 by thirteen-year-old Justin Schiller (a founding member as well of the LCSNA 17 years later), it became the International Wizard of Oz Club in 1959. Gardner served on its first Board of Directors.

Martin Gardner: *Ave Atque Vale*

MORTON N. COHEN

We recognize Charles Dodgson as a multifaceted personality with intellectual tentacles reaching out in many directions. Martin Gardner easily matches Dodgson's outreach and in fact goes further, into areas that Dodgson did not contemplate. The record of the two lives overlap in many ways: both men devoted much of their energies to mathematics and logic, and their flights of fancy sent them soaring into the realms of storytelling and poetry. They both wrote on a large variety of subjects, not as dabblers but as committed scholars, seriously but always with an undercurrent of wit. They tackled the world's mysteries rationally, but with an engaging, enchanting whimsy.

I first met Martin on Saturday, January 12, 1974, at the founding meeting of the Lewis Carroll Society of North America, held at the Nassau Inn in

Long considered the doyen of Carroll studies, LCSNA founding member Morton N. Cohen has written or edited, among innumerable scholarly articles and many other books, The Letters of Lewis Carroll *(1979),* Lewis Carroll and the House of Macmillan *(1987),* Lewis Carroll: Interviews and Recollections *(1989),* Lewis Carroll and His Illustrators *(2003),* Lewis Carroll: A Biography *(1995), and* Reflections in a Looking Glass: A Centennial Celebration of Lewis Carroll, Photographer *(1998). The Modern Language Association set up the biennial Morton N. Cohen Award for a Distinguished Edition of Letters in 1989, and he was appointed a Fellow of the Royal Society of Literature in 1996.*

Princeton. We were a group of some fifteen strangers, brought together by rare book collector and dealer Stan Marx to establish a society on this side of the Atlantic, inspired by the Lewis Carroll Society that had been established in England five years prior. A few of us had already published books or articles on Lewis Carroll, and this was a remarkable opportunity to meet each other. In addition to Martin and Stan, among others there were Carroll biographer Florence Becker Lennon and Alexander Wainwright, the Princeton University librarian who kindly arranged for the occasion an exhibit of some of the Lewis Carroll treasures in the Parrish Collection. Looking through my files, I find an article, "Merrily We Roll Along," which I wrote for *Thimbles and Hope*, the publication edited by August Imholtz, Jr., on the occasion of the Society's twenty-fifth anniversary in 1999, with some memories of that initial meeting. An excerpt or two seeks to catch the prevalent mood of the occasion:

"We gathered there," it reads, "'round a wooden table, sat on wooden chairs, looked upon one another, introduced ourselves, shook hands, uttered civilities. Some, as strangers to most of the others, felt slightly awkward, somewhat embarrassed. What could we say to one another? Did we have anything [else] in common?" But that one shared interest was apparently enough: it "alone quashed the hesitancies, dispelled the reticence, cleared the air, and, lo! We soon found ourselves speaking to one another, framing ideas, elaborating upon elementary notions, adding suggestion to suggestion, all of us fired by the hope that something far-reaching, possibly permanent, would come of this meeting."

" . . . Martin Gardner," the text continues, "had just published a semi-autobiographical novel, entitled *The Flight of Peter Fromm*, which Martin Levin in *The New York Times* called, 'a brilliant illuminating metaphysical novel that employs ideas as adversaries and translates them into human dilemmas.' And even as I am writing this piece, a parcel has arrived from Martin containing his latest novel, *Visitors from Oz*, containing (Eureka!) five chapters about Wonderland."

Martin reappeared at some of the subsequent meetings of the Society. I recall one executive meeting that I had arranged at the Graduate Center of the City University of New York, where I was then teaching, which Martin and his wife attended. Not long after that, he was instrumental in getting Norman Armour, Jr., whom he knew, the owner of the galley proof of

the wasp chapter of *Looking-Glass* that John Tenniel refused to illustrate, to allow the Society to publish the missing chapter. In fact, Martin wrote the introduction to the Society's handsome booklet of *The Wasp in the Wig*.

Of course we corresponded through the years, and Martin was especially helpful to me when I ran into mathematical or logic problems while editing the Carroll letters. He wrote a dust jacket blurb for my Carroll biography, and I reviewed his *Definitive Annotated Alice* in the journal *Victorian Studies* and in the English Society's *Lewis Carroll Review*. Martin came once to visit me in my apartment on Barrow Street, when he wanted to check some material in the photocopy of Dodgson's diaries that I have. It was a special pleasure talking to this calm and gentle man about many things over the English version of a cream tea that I had prepared. Yes, he was soft-spoken, amiable, and shy—he refused repeatedly to give any lectures—but when he got involved in controversies as he did over one about black holes in the universe that raged in the *New York Review of Books* for a time, his prose was fiery, and it was not long ago that he wrote a sharply worded denunciation of the Karoline Leach revisionist cabal in the *Knight Letter* (No. 75, Summer 2005). Martin was a relentless debunker of pseudos and cranks.

Indeed, it was the great range and diversity of his interests that marked him the Baconian man of our time. In his book of 47 essays, *The Night is Large: Collected Essays 1938–1995*, he arranged his intellectual interests under these headings: Part I, Physical Science; Part II, Social Science; Part III, Pseudoscience; Part IV, Mathematics; Part V, The Arts; Part VI, Philosophy; and Part VII, Religion. In his most recent book, *The Jinn from Hyperspace and Other Scribblings—Both Serious and Whimsical* (2008), he assembled 36 essays, arranged under these headings: Part One, Science, Math, and Baloney; Part Two, Literature; Part Three, L. Frank Baum; and Part Four, Lewis Carroll.

This last book of his has a special meaning for me. Since I retired from university teaching, I have been able to escape New York's brutal winters by spending three of the worst months in Puerto Rico. When I returned to New York in March of 2008, I confronted the customary accumulation of unopened mail. As I went through the morass, I came upon a parcel that obviously contained a book. It was this last book of Martin's. I was delighted, of course, to have it, but imagine my reaction when I opened it to find that the book was dedicated to me. On the flyleaf Martin wrote in his own hand: "I'm told it is customary to get permission to dedicate, but I wanted this to be a surprise. All best wishes to you for 2008. Martin."

Martin lived a long, full, and successful life, and his accomplishments will shine as beacons to new minds of oncoming generations, to people who will seek to understand all manners of meanings and truths. He will inspire them as he has inspired us.

Martin Gardner, an Appreciation

SELWYN GOODACRE

In 1960, I was a penniless third-year medical student, but had been collecting Lewis Carroll for about five years—largely magazine articles and whatever biography I could find, as funds were so meager. I heard about a new book, *The Annotated Alice*, having been published in America. I ordered a copy from Hudson's Bookshop in Birmingham. Its price, $10, at that time seemed a huge amount. It took a while to order, and I asked my then-fiancée Janet (we've been together now for over forty years!) if she could possibly pick it up for me.

What an incredible book! I can remember my amazement and astonishment at the sheer opulence of the book—so large, with thick laid paper, the text in large type with annotations alongside, all so elegantly presented. It all seemed to make my ambition to concentrate on collecting Lewis Carroll as a lifetime project so much more worthy and worthwhile.

Curiously, it took a while before *The Annotated Alice* was published in England—by Anthony Blond (though it still bears the 1960 date). It was towards the end of 1964 before reviews began to appear—but it then took

Selwyn Goodacre, a retired physician, noted collector, and former chairman of the Lewis Carroll Society (UK) and editor of its journal, Jabberwocky, *for 23 years, has spoken and published widely on textual, bibliographic, theological, and medical aspects of Carroll's life and work. He was the editor of the Pennyroyal* Alices; *his most recent book is* The Illustrated Editions of the Hunting of the Snark *from Inky Parrot Press.*

off like a storm. An authoritative review in *The Times Literary Supplement* was followed by an extended series of "Letters to the Editor" that continued for nine months. It is difficult in hindsight to appreciate just what a breakthrough this book was. Children's books were not thought worthy of that degree of attention in those days. One could say that it heralded the renewed academic approach to the life and works of Lewis Carroll, paving the way for the publication of *The Letters of Lewis Carroll* some years later, and indeed, the formation of Lewis Carroll Societies the world over.

The concept of "annotated texts" of great works of literature may well be said to have been inspired by Martin's pioneering work on the *Alice* books. Now they are a regular feature of new publishing, from Jane Austen to Kenneth Grahame.

My admiration for this book has not abated. Gardner encouraged and welcomed new ideas from readers, and as a result the book went into numerous reprints (see the Bibliography, p. 207), spawning sequels and "authoritative" re-issues. Martin loved the whole process. I plucked up courage to write to him myself and offer some of my own thoughts—which, in that wonderfully generous nature of his, he gladly incorporated into those later editions. My daughter was taking ballet lessons at the time and remarked to me that when Alice said the Lobster's toes demonstrated "the first position in dancing" Tenniel's picture indeed shows just that. And Martin put that thought in a later edition. I cherish to this day his acknowledgement to me in the preface to *More Annotated Alice*.

In the early 1990s, thanks to Charlie and Stephanie Lovett, I had the privilege of meeting him in North Carolina, and he was as charming and delightful as he could possibly be. It was an emotional moment for me to meet one of the great Carrollian scholars for the first time.

I may not have shared all his opinions of the *Alice* books—he had problems coming to terms with the fact that children did enjoy them; and indeed that that was their primary purpose. He set out his views in his definitive article, "A Child's Garden of Bewilderment" (*Saturday Review*, July 17, 1965), an article whose conclusions I totally disagreed with. But he loved to correspond with enthusiastic readers of his work.

In 1972 he went on to annotate "The Hunting of the Snark," another work that went into many manifestations. In 1981, William Kaufmann, Inc. of Los Altos, California, in cooperation with Bryn Mawr College Library, published

an ambitious project celebrating the poem, including Martin's annotations, as well as full accounts of its genesis. Martin recalled that I had issued a little booklet, *The Listing of the Snark* (in 1974), and suggested to the publisher that I revise it for inclusion in the Kaufmann edition. Such a generous thought, and I am hugely grateful to his memory for his support and encouragement.

Martin was a wonderful correspondent, always answering very fully any comments one had made to him. For some reason, in one letter to me, he included a clipping of a letter he had sent to the journal *Norman [Oklahoma] Transcript* about the truth of evolution. I responded with some enthusiasm, as I have never been able to understand why some fellow Christians want to deny evolution: if man can evolve new breeds of dog by selective breeding, it seems simply perverse to suggest that God "stopped" creating at Eden.

At about that time I mentioned that my son, Mark, was a professor of theology at Duke University. In a letter (January 25, 2006) commenting on this, he said "I guess we hold similar views on intelligent design. As a philosophical theist I see no reason why God can't use the combination of chance (mutations) and natural selection as a method of causing cosmological and biological evolution. If the laws are created and upheld by God, there is no more need for special interjections along the way than a need for God to keep adjusting planetary orbits, as Newton believed."

Wise words, and so beautifully expressed.

As well, I loved his various books explaining physics and mathematics to the common man, and his delight in the quirks of conjuring, "magic," and illusion. I once shared with him the foolproof way a general Medical Practitioner can always predict the sex of an unborn child correctly: Tell the mother it will be a boy, but write a note saying it will be a girl, signed and dated. If it's a boy, the mom is delighted; if a girl, tell the mom she heard it wrongly—and prove it with the signed paper. He loved it.

Even in his nineties he was full of ideas. In a letter to me just after he reached the age of ninety, he was full of ongoing projects: "I am updating some of my old, out-of-print books for new editions, and writing occasional book reviews, so I keep busy." He mentioned he was writing a supplement of "new notes for my *Annotated Alice* that will be published in the next issue of *Knight Letter*."

In 2006 I nervously sent him a draft of one chapter from my projected "Textual Commentary on *Alice's Adventures in Wonderland*," seeking his approval, though fearing that he might be a little offended that I was daring to tread on his ground. But not a bit of it. He was enthusiastic: "I'll be eagerly awaiting the book." I am only sad that he died before it was published.

Martin Gardner has done great work in Carroll scholarship; we shall always be in his debt. We must never underestimate his invaluable, unique, and pioneering contributions.

Perhaps a Few Wise Words: A Tribute to Gartin Mardner

EDWARD GUILIANO

Causalities are tricky things to chart, but Martin Gardner deserves much credit for the popularity and acceptance of Lewis Carroll and his *Alice* books in the academy today and well beyond in our global culture.

It all began, of course, with a boat ride that took place on a golden afternoon on July 4, 1862, even if Martin Gardner later revealed to a wide audience that it was a cloudy, damp day. We learned about the weather and much, much more with the publication of his *The Annotated Alice* in 1960, a seminal date in Carroll studies and in the broadening appreciation and fascination of adults with the *Alices*. Thanks to Carroll and Martin, it became almost reverse-chic then chic-chic and an intellectual compliment to quote *Alice* or to give someone a copy of *Alice* or books or articles about her.

In 1862, there was no *Alice in Wonderland* for Martin to annotate. And no Martin. There was a reading public that liked to read fiction to "relax

Dr. Edward Guiliano, a founding member and past president of the LCSNA, is President and CEO of the New York Institute of Technology (NYIT) and recognized around the world as an eloquent spokesman and advocate for global higher education, sustainability, and emerging technologies. A renowned scholar and widely published author on Victorian literature, he has edited the essay collections Lewis Carroll Observed *(1976),* Lewis Carroll: A Celebration *(1981), and* Soaring with the Dodo *(1992); compiled* Lewis Carroll: An Annotated International Bibliography *(1977); edited* The Complete Illustrated Works of Lewis Carroll *(1982); and was instrumental in the publication of* The Wasp in a Wig.

the mind." The Victorians also read to learn and for inspiration. They liked moralized stories and sermons and religious literature of all kinds. The Reverend Dodgson's library bears witness to this.

No one studied literature as we think of it. Indeed, studying literature meant reading the classics in the original Latin or Greek. Contemporary literature in English simply was not taught at universities until the twentieth century, and then first in America. In the 1890s studying English, and only later American, literature as a separate subject was a new thing. Almost all courses, especially in England, were focused on philology and on the survey of the history of the English language and literature pre-1800. During that *fin-de-siècle* period at the close of Dodgson's life, however, American literature professors were beginning to expand their focus to thematic topics and literary movements.

We know from Dodgson's library and elsewhere that he read Dickens, and that he owned all of his novels in first edition. But not even the 1862 Dickens was the day's best-selling author. The public was reading Mrs. Oliphant, who had two best sellers that year, *The Doctor's Family* and *The Last of the Mortimers.* (I confess to not having read them.) And then, of course, there was Margaret Elizabeth Braddons' *Lady Audley's Secret.*

Lewis Carroll has been read consistently since 1866 and, needless to say, has been alive in the popular culture continuously to this day. But he was not *studied.* Literary and critical reputations rise and fall. The reputations of many, many of Carroll's contemporaries have declined. Those poets and essayists as well as novelists who were revered in their day are less so today. Even the inimitable one, the great Dickens, fell in favor the first half of the last century. But he, like Carroll and Conan Doyle, created characters and stories more famous than their authors, and their names and works have survived in movies and plays and the various communication accoutrements of a globalized modern world. As their posthumous productivity proves, they have earned continued respect and admiration, especially Carroll, whose audience grew far beyond children.

Until Martin Gardner, Carroll was not a suitable university-approved author and subject. He was not on reading lists. Papers on his works were not delivered at learned conferences.

After Carroll's death, as is the custom with famous writers, there was a spurt of biographies, letters, bibliographies, and the like. But it really was not till 1935 that Carroll was included in a major scholarly study, William

Empson's "The Child as Swain" essay in his book *Some Versions of Pastoral*. Then there was Elizabeth Sewell's still somewhat incomprehensible *The Field of Nonsense* in 1952. And Phyllis Greenacre's psychoanalytic semi-biographical study of Swift and Carroll, published in 1955. But they were somewhat tangential exceptions to the rule. Sure, there were other essays of merit over the 60-odd years after Carroll's death, but not sustained academic or critical interest and commentary as we now think of it. Carroll was not included in either the first edition of *Victorian Fiction: a Guide to Research* in 1964 or the second edition in 1978. Yet by 1980, academic interest in Carroll was easily eclipsing some of the most venerable Victorian authors . . . and I don't mean Mrs. Oliphant.

The year 1960 was a time when American higher education was expanding rapidly; more than half of the colleges and universities that exist in America today were founded after 1955. And in the 1960s the American academic complex began a scholarly publications race. Still not much Carroll, but Martin Gardner was opening up the world of Carroll's *Alices* for us, in a sense showing us there is no such thing as children's literature, as adults can take away meaningful insights and pleasures from what was reductively called "children's books." He brought us to a work of art and enabled us to see more in it, take away more, and delight in the brilliance of the games, logic, story, comedy, and language. He opened up adult themes and universal relevancies.

Et voilà. Today Carroll's place in the academy is solid and stronger, as I have noted, than many, forgotten Victorian poets in particular, as well as novelists. If one wants to fix a date when Carroll became acceptable in the learned halls of academia, it was 1966, when the journal of the venerable Modern Language Association, *PMLA*, printed Donald Rackin's essay, "Alice's Journey to the End of Night." The copy text Rackin used for all his citations from *Alice in Wonderland*? Martin Gardner's *The Annotated Alice*.

And this emergence and acceptance points, for me, to the curious mind with eclectic and broad tastes and interests that wrote *The Annotated Alice*.

I met Martin Gardner in Princeton in 1974 at the inaugural meeting of the Lewis Carroll Society of North America, an organization founded in part to promote serious studies of Carroll's life and art. Martin was the oldest person there, and I was a young Ph.D. student. And for the next seven or eight years, until he moved away from New York, he was a great friend to me and to the Society.

He was shy and had some quirky ideas, sometimes not being able to distinguish the relative merits of a small idea or example from a big one. He had diverse intellectual passions that were a joy to be around. And he was kind and generous. Despite his *Scientific American* column, he was a freelance writer who hustled all the time. He took me with him to Dover Publications to discuss reprints. He took me to Clarkson N. Potter, Inc., to meet Clark Potter and pitch *Lewis Carroll Observed*. He came by our apartment and talked with me there and on the phone again and again about ideas and projects.

For *Lewis Carroll Observed*, I found myself being his editor on what we can now say is one of the lesser essays in the book, but an oh-so-Martin Gardner essay: on the source upon which "Speak Roughly" is a parody. And there I was, young and innocent, editing the prolific author, trying hard to make the best essay we could. Awkward for me, certainly, yet he could not have been more appreciative or easy to work with. It taught me a lot. He was a great man, and I was just one among many he graciously helped to grow.

During one of our many conversations, he shared with me that he found the *Alice* books frightening when he was a child but fascinating as an adult. I will never forget that because that is how I felt, too, and we talked about it, and he had reasoned out all the whys, and it was a liberating discussion.

There are many Martin Gardner stories I could tell in tribute; here is the one on how *The Wasp in a Wig* came into being, because it illustrates that only because of Martin was it published by the Lewis Carroll Society of North America—with later commercial editions by Clarkson N. Potter in America, and Macmillan London Limited in England. At the time, I was the publications committee chair of the LCSNA and once again found myself working with and editing Martin, who wrote the preface, introduction, and annotations.

Back in 1898, Stuart Dodgson Collingwood, in his *Life and Letters of Lewis Carroll,* mentioned that *Through the Looking-Glass* originally contained a "wasp in a wig" episode, but it was dropped when Tenniel wrote that it wasn't up to the rest of the book, and went on to say, "I can't see my way to a picture. If you want to shorten the book, I can't help thinking—with all submission—that *this* is your opportunity." Collingwood reproduces the

Tenniel letter in facsimile in his book, and Martin Gardner reprints it in *The Annotated Alice*. For the next 77 years, that is essentially all anyone knew about the episode. No one had ever acknowledged having seen it, and there were no other references to it. It doesn't appear in the list of Dodgson's possessions that were auctioned after his death. Everyone assumed that is was destroyed and agreed it was a great pity that we could not have a glimpse of such a great curiosity.

Then, without the least bit of fanfare, at Sotheby Parke Bernet, in London on June 3, 1974, galley proofs of the episode with hand corrections came up for auction. None of the Carroll collectors and scholars in England had had the opportunity to read the galleys so to at least give us some idea of the contents and merits. The galleys were sold, at a very modest price, to a rare book dealer in New York City.

As soon as I learned that the galleys had come to New York, I contacted the dealer in hopes of viewing them. He no longer had them. He had purchased them in London for a wealthy client "as an investment." The dealer, as is the standard practice, would not tell me the name of the galleys' new owner, although he did pass along several letters that I wrote. None were answered. It began to look as if the galleys might be lost for another hundred years. I discussed the situation with Martin and some members of the LCSNA's publications committee, and we decided Martin should write a letter, as he was so well known and the owner might respond. By then I had a clue to the owner's identity, and with the help of another friend, we obtained the address, and Martin wrote the letter. This time the owner, Norman Armour, Jr., responded with a telephone call to Martin, who was subsequently invited to Amour's apartment to view the galleys, which he assured me afterwards deserved to be published on merit as well as on historical significance.

He and I then began a modest campaign to gain Armour's permission to publish the galleys. On December 9, 1976, Martin wrote to him, "Dear Norm: As a sort of Christmas gift I've put in the mail for you a copy of *Lewis Carroll Observed*. I have an essay in it on the mystery of who wrote the poem that Carroll parodied in 'Speak Roughly.'" He went on to explain that a publication of a *More Annotated Alice* is not likely before 1980:

> Now 1980 is more than three years away, and it seems to me that this is too long for Carrollians to wait to see the lost

> episode. What do you think of the following plan? Ed Guiliano is editing a series of chapbooks, each dealing with some aspect of Carroll. I enclose the first one. Others will be uniform in format. . . I was asked if I would write a chapbook about the lost episode, why it was dropped, how it was preserved, and how it finally came into your hands. . . . Publication of the chapbook would be a news event that I believe would get good coverage in news magazines and newspapers. . . . If you give permission, we would start work at once, with publication of the chapbook next year.

Amour was a lovely gentleman who was neither a serious book collector nor a literary-minded individual, and at first had little idea of what he owned. He had a mild interest in some classic children's literature, but he was most interested in the galleys as a way to beat inflation and taxes. At first he was adamantly opposed to our proposed project because he was concerned that publication would decrease the galleys' value. With a bit more of coaxing (now I was involved in telephoning him, too), Armour agreed, and the Dodgson estate agreed, and *The Wasp in a Wig* was published in 1977. Neither Martin Gardner nor Norman Amour benefited financially in any way from the publication; they were characteristically generous with their time and Martin with his extraordinary expertise. When Armour passed away, the galleys went to his heirs, and to the best of my knowledge are likely to still be lodged in the bank vault from which we "borrowed" them to be photographed for publication.

Martin's role in their publication is testimony to his munificence and accomplishments. Gartin Mardner—as he sometimes signed his name on book dedications—inspired, enriched, and entertained many lives. When it came to Carroll, he not only agreed that "*Everybody* has won, and *all* must have prizes," he gave them to us.

The Annotated Martin Gardner

PETER E. HANFF

To celebrate the fiftieth anniversary of the founding of The International Wizard of Oz Club, I proposed to Martin Gardner, a founding member of the Club, that he participate in an interview with Club member Kevin Thomas. The interview was conducted in Martin's apartment in Tulsa, Oklahoma, in the spring of 2007, and captured on digital videotape. I completed a transcription in June, 2010.

In 1956, Roland Baughman, head of the Department of Special Collections at the Columbia University Libraries, created a major exhibition of L. Frank Baum to celebrate Baum's centennial. This was the first academic library tribute to Baum and revealed to the viewers (and the world through a published catalogue of the exhibition) what a rich and sophisticated array of works Baum had produced in a relatively brief writing career spanning the years 1897 to 1919.

Peter E. Hanff is the Deputy Director of the Bancroft Library at the University of California, Berkeley; he has made significant contributions to the scholarship on L. Frank Baum and Oz, perhaps most notably his collaboration with Douglas G. Greene on the Bibliographia Oziana. *A leading authority on rare books and special collections, he has at two different times served as president of The International Wizard of Oz Club, and is the director of the Winkie Convention.*

Beyond his children's classic, *The Wonderful Wizard of Oz*, originally published in 1900, Baum published about sixty books. Most were for children or adolescent readers, but he also published three novels for adults. Much of Baum's work was published under pseudonyms to avoid competing with his annual output of a major original children's book each year. Few members of the American public in later decades were aware of Baum's prolific output, but among those few were Roland Baughman and Martin Gardner.

The mid-1950s marked a low point in Baum's reputation, but also a turning point, for through the efforts of Baughman, Gardner, and other like-minded enthusiasts, a renaissance of appreciation began to evolve. In 1955 Martin Gardner published a two-part essay in *The Magazine of Fantasy and Science Fiction*, assessing Baum's innovative and imaginative contributions to children's literature. That essay, complemented with a biographical sketch by Russel B. Nye, served as the scholarly part of *The Wizard of Oz and Who He Was*, published by the Michigan State University Press in 1957. Gardner and Nye even made sure that some minor (and non-authorial) changes made by Bobbs-Merrill in 1903 to Baum's text of *The Wonderful Wizard of Oz* were reversed.

In 1957, thirteen-year-old Justin G. Schiller launched The International Wizard of Oz Club with a founding membership of just sixteen, including Martin Gardner. Thirty years later membership peaked at three thousand. Clearly Martin Gardner's own enthusiasm for Baum and Oz was shared by many.

Then, in 1959, Martin Gardner published *The Annotated Alice* through Clarkson Potter. Certainly parallels between Alice and Dorothy were not lost on Gardner, although I think he always recognized that each character reflected a specific cultural context.

Readers of this festschrift may enjoy reading some of Martin Gardner's memories and comments in his own words. They may be surprised by his initial perspective on Alice:

KEVIN THOMAS (KT): Please tell us something about yourself.

MARTIN GARDNER (MG): Well, I grew up in Tulsa. And I grew up on the Oz books. In fact, I learned how to read by looking over my mother's shoulder when she read *The Wizard of Oz* aloud to me. And that made things very embarrassing for me in the first grade, because the teacher would hold up sheets that said *cat* and *dog* and I would be the first to call out the word. She finally had to tell me to shut up so the other students could participate.

But I loved the Oz books. I read all of them when I was a small boy. And I persuaded my parents to buy all Baum's non-Oz books. Some of them I enjoyed even more than the Oz books. I think some of his non-Oz fantasies were even better written than some of his Oz books. I'm particularly fond of *The Magical Monarch of Mo*, which is one of his earliest books, and *John Dough and the Cherub* I thought was a great book. [*Peter Hanff comments:* The Magical Monarch of Mo, *published in 1903 by Bobbs-Merrill, was a lightly revised version of* A New Wonderland, *originally published in 1900 in New York by R. H. Russell. In his personal copy of* A New Wonderland, *Baum wrote out the history of the text, explaining that it was the first book for children he wrote—he copyrighted it in 1896 as* Adventures in Phuniland, *and it is probably no coincidence that both titles of the work remind us of Adventures in Wonderland.*]

I think *Sky Island* [1912] was one of his masterpieces.

KT: Tell me some of what you did throughout your life.

MG: Well, I started out . . . my first job was as a reporter on the Tulsa *Tribune*, a paper that doesn't exist anymore. That was early in the Depression, and I think I made $15 a week. After I graduated from the University of Chicago, I got a job in the press-relations office of the University of Chicago. I was there until I listed in the Navy during World War II. I did a four-year hitch in the Navy on the Destroyer Escort in the Atlantic.

When I got out of the service I made my first sale to *Esquire* magazine. It was a short story.

For about a year after World War II, I was supported by sales to *Esquire* of short fiction.

I finally decided that the place budding writers should be would be in New York City, so I moved to New York City and got a job as assistant editor of *Humpty Dumpty*, a magazine for very young children. I did that for eight years. And I sold an article to *Scientific American* and that was my big break.

They published it as an article and Gerry Piel, who was the publisher, called me into his office and asked me if there was enough similar material to make a regular column. I jumped at the chance, and said, yes, I thought there was.

I rushed around New York in the old-book-store region and bought all the books on recreational mathematics I could find, and that was the start of my column, which I did for twenty-five years. That was my one big break, because with those columns, with the release from *Scientific American* [to] publish them as books, I got fifteen books out of just those columns alone.

Since World War II I've been a freelance writer.

KT: One of your books, as we all know, was *The Annotated Alice*, so tell us about that please.

MG: Well, I was not fond of Lewis Carroll's *Alice* books when I was a small boy. They sort of frightened me. It was not until I was in my twenties when I began to read the *Alice* books. I came to be tremendously interested in Lewis Carroll, who was, by the way, a professor of mathematics. It occurred to me that the *Alice* books were filled with so many hidden jokes and hidden puns and subtleties on events that were recognized by children in England [in their] time but [lost on modern readers]—without a footnote you missed a lot of the humor in the books. I proposed to an editor that they find someone to do an annotated edition of *Alice* and I mentioned Bertrand Russell, who was a person who was fond of the *Alice* books. Several editors actually did write to Russell, and he turned them down. He was not interested.

One young editor named Clark[son N.] Potter, when I told him the story about Bertrand Russell, said why don't *you* do it? And I said I'd give it a try, so that's how I happened to write it.

Dr. Matrix in Oz

MICHAEL PATRICK HEARN

"Ten Commandments . . . can't mend most men."

– *The Incredible Dr. Matrix*, 1976

Martin Gardner was my literary godfather and the most generous man I have ever known. I owe him everything. He was both incomparable and indefinable. "Martin Gardner's contribution to contemporary intellectual culture is unique—in its range, its insight, and understanding of hard questions that matter," Noam Chomsky acknowledged. Stephen Jay Gould considered him to be "the single brightest beacon defending rationality and good science" of the second part of the twentieth century. "If Edmund Wilson was, as they say, the principal American man of letters in our time," declared Pulitzer Prize–winning critic Michael Dirda in *The Washington Post* in 1996, "then Martin

Michael Patrick Hearn has specialized in juvenile literature since the publication of his first book, The Annotated Wizard of Oz. *His many other works include* The Annotated Christmas Carol, The Annotated Huckleberry Finn, The Victorian Fairy Tale Book, Myth, Magic and Mystery: One Hundred Years of American Children's Book Illustration, W. W. Denslow: The Other Wizard of Oz, From the Silver Age to Stalin: Russian Children's Book Illustration, *and* The Porcelain Cat, *illustrated by two-time Caldecott Medal winners Leo and Diane Dillon. He is currently working on* The Annotated Edgar Allan Poe, *and one day he may actually finish his long overdue biography of L. Frank Baum.*

Gardner must be our leading man of numbers." *The Guardian* called him "the sage of skepticism." He was "the Indiana Jones of the Western intellectual tradition" to *The Los Angeles Times*. Mathematician Rudy Rucker said he was "the dean of puzzledom, the ringmaster of mathematical games." Prof. Clarence Brown of Princeton University thought he was "arguably the most cerebral Middlebrow of the twentieth century." I just called him my friend.

FROM TULSA TO WONDERLAND

Martin Gardner, journalist, mathemagician, literary critic, philosopher, debunker, polymath, Ozmapolitan, Carrollian, was fortunate to spend most of his ninety-five years doing exactly what he wanted. "I just play all the time," he told *Skeptical Inquirer* in 1998, "and am fortunate enough to get paid for it." Life was not always that easy. Born in Tulsa, Oklahoma on October 21, 1914, he was the oldest of three children. He had a brother, Jim, and then a sister, Judy. With a Ph.D. in geology, his father was a prosperous "wildcatter" who owned a small oil business. His mother was a former Montessori kindergarten teacher. His father tolerated her devout Methodism that troubled their oldest son for the majority of his life. Martin's struggle between the teachings of God and the proofs of science fueled much of his later writing. His father once admitted that Martin may have been the sanest one in the family. Another thing that troubled Martin all his life was that Tulsa was the location of one of the worst race riots in American history. In 1921, Greenwood, the Black district, was burned by marauding whites returned from the war. Martin was not proud of his parents' racist attitudes during his childhood, but at least his father gave generously to rebuild a Black church that had been destroyed.

The child learned how to read by looking over his mother's shoulder while she read him *The Wizard of Oz*. Of course, it caused problems in the first grade. When the teacher held up language cards in class, he was always the first to shout out the words until she finally told him to be quiet and let the others try. He went through all of Baum's Oz Books and begged his parents for his other works. He particularly enjoyed *The Magical Monarch of Mo* and *John Dough and the Cherub*, and agreed with Baum that *Sky Island* was one of his masterpieces. Curiously Martin did *not* like Lewis Carroll's *Alice* books when a little boy: they frightened him. Yet he never lost his early love for Oz. The stories made such an impact on the child that he could not look at a bowl of bright green Jell-O without thinking of the Emerald City, or see a little girl in blue without being reminded of a Munchkin. More importantly, *The Wizard of Oz* provided a template for his life's work. He became a professional Dorothy who left the Great American Prairie to expose the humbug wizards behind their screens of deception and douse the intellectual witches with good, clean, sensible buckets of logic. They may have not entirely melted away, but he left them all wet.

The child was indeed father to the man, as Martin developed many of his lifelong interests while growing up in Tulsa. His father introduced him to science through astronomy and geology. The boy was an inveterate collector of butterflies, house keys, match boxes, and stamps, but he gave them all up for his new favorite hobby, magic. His father had taught him a few tricks and he met some local magicians when he was in high school. He acquired, along with a Mysto Magic set, all the mechanical puzzles that could be ordered by mail. "My main interest in magic is because it arouses a sense of wonder about the natural world," he explained many years later. "The universe is almost like a huge magic trick and scientists are trying to figure out how it does what it does."

What first hooked him on recreational mathematics was another gift from his father, Sam Lloyd's *Cyclopedia of Puzzles* (1914). The lavishly illustrated children's magazine *John Martin's Book*, particularly the "Peter Puzzlemaker" page drawn by the art director, George Carlson, also fascinated the boy. Another important influence on him was the popular monthly *Science and Invention*, which combined real science with wild speculation. *Amazing Stories* fueled his appetite for science fiction. Other interests included Sherlock Holmes stories, Tom Mix movies, and Western serials. He played a lot of chess, but finally gave it up because he found it left him little time to do anything else. He was editor of his elementary school newspaper. His family had a tennis court, so he played a lot of tennis. He specialized in the horizontal bar on the gymnastics and tumbling team and fell head over heels for "that elegantly choreographed dance called baseball." (He thought boxing and bullfighting should be outlawed.) Saxophone lessons did not go far, but somehow Martin, like Marlene Dietrich, learned how to play the musical saw. Not well, mind you.

High school bored him. He called it his "four-year prison." He complained so much one morning his father finally gave in and let him play hooky. His dad wrote the school, "Please excuse Martin's absence yesterday. He was sick at home with the gripes." He did well in geometry and physics but flunked Latin. He admitted he was never any good at foreign languages. "Biographical history, as taught in our public schools," he wrote in *The New York Times* in 1976, "is still largely a history of boneheads: ridiculous kings and queens, paranoid political leaders, compulsive voyagers, ignorant generals—the flotsam and jetsam of historical currents. The men who radically altered history, the great scientists and mathematicians, are seldom mentioned, if at all." He joined the Press Club and wrote poems for school publications. He broke into print out of school at age sixteen in 1930 through snippets in *Science and Invention* and *The Sphinx*, a magazine for magicians. He must have been a doodler because his Senior yearbook described him as "an able cartoonist with an adept mind for science."

Certain he was an atheist when he entered high school, he defiantly did not bow his head during prayer in class. Then a counselor at Camp Mishawaka in Minnesota converted him to Fundamentalism; for a year Martin actually attended a Seventh-Day Adventist Church. After reading *The New Geology* (1923) by the Creationist George McCready Price, the young believer was certain Armageddon was at hand and even thought "evolution was a satanic myth" until his high school geology teacher set him straight. That realization set in motion his lifelong mission to debunk false science.

He longed to be a physicist, but when he applied to CalTech, where Nobel Prize–winner Robert A. Millikan was chief physicist, he was informed that he needed to complete two years at a liberal arts college first. He entered the class of 1936 at the University of Chicago a Fundamentalist Christian and turned from physics to philosophy to determine his own beliefs. "Philosophy gives one an excuse to dabble in everything," he admitted to fellow science writer Kendrick Frazier in 1998. Except for auditing geology, the only science instruction he received at the University of Chicago was in the survey course Physical Science. He eventually lost all faith in Christianity at college. Nevertheless he wrestled with religion throughout his adult life.

It was an exciting time to be going to the university. Despite his running the campus like a dictator, Robert Hutchins, the youngest president of any major college in the country, introduced a radical new plan that gave the students incredible freedom in choosing their courses. Martin took classes taught by writers Norman Maclean and Thornton Wilder and attended lectures given by Socialists Max Eastman and Norman Thomas, another hero. He could audit any course he wanted (of course without getting credit). The only requirements were four survey courses in the physical sciences, biological sciences, social science, and humanities. Hutchins brought to the university the Aristotelian philosopher Mortimer J. Adler, whom Martin never liked. Both Hutchins and Adler supported Columbia University's Great Books Movement, which Martin never really understood. Why did they include *Little Dorrit* rather than almost any other Dickens novel? They should have selected *Ulysses* rather than *Portrait of the Artist as a Young Man*. Why not *Alice's Adventures in Wonderland* or *The Wizard of Oz*? Martin preferred them to something like *Middlemarch*.

Adler was a character with a highly inflated ego. Though born an Orthodox Jew, he toyed with Catholicism and went so far as to say that the Church had every right to execute heretics. A joke at the time was that the University of Chicago was a Baptist school where Jewish professors taught Catholic theology. Martin audaciously wrote to *The New Republic*, begging readers to pray to God that Adler, like Chesterton, would convert, which he did just shortly before he died.

While at college, Martin wrote with some regularity for *The Sphinx* and issued a series of pamphlets of magic tricks. He gathered much of his material from a group of professional magicians who held a weekly roundtable meeting in Chicago. He drew caricatures for *The Phoenix*, the college humor magazine, and published the very first article on puzzles to appear in *Hobbies Magazine*, in September 1934. During the Christmas season, he earned a little extra money by performing magic tricks to promote Gilbert magic sets at Marshall Field's Department Store. After graduating with a bachelor's degree in philosophy in 1936, he took a job as assistant oil editor of the *Tulsa Times* for $15 a week, but the work was so dreary that after a year he returned to Chicago, where he joined the Press Relations staff of the University of Chicago, and the Chicago Relief Administration. He was also Contributing Editor to the school's literary journal, *Chicago Magazine*, and organized an exhibition of books by alumni to commemorate the school's fiftieth anniversary.

Life was a struggle then. In October 1940, the student magazine *Pulse* floridly described him as "a slim, middling man with a thin face saturnined by jutting, jetted eyebrows and spading chin; his simian stride and posture is contrasted by his gentility and fluid deftness of his hands." He was indifferent to money and possessions, "a Robinson Crusoe by choice, divesting himself of all material things." He was known to be "periodically down to his last five dollars, facing eviction from the Homestead Hotel, and triumphantly turning up, Desperate Desmond fashion,* with fifty or a hundred dollars at the eleventh hours—the result of having sold an idea for a magic trick or a sales promotion angle to one of a half-dozen companies who look to him for specialties." One may never know exactly all that he did during those early Chicago days. *Pulse* reported that he succeeded in selling "ideas for booklets on paper-cutting and other tricks, 'pitchmen's' novelties, straight magic and card tricks, and occasional dabblings in writings here and there have made him even more well known as an 'idea' man for small novelty houses and children's book publishers." As the author of *Match-Ic* (1935) and other pamphlets, he sent *Life* magazine (March 3, 1941) a new trick for making a "Nazi cross" without breaking the matches: "Stick four matches in his ear and light them with the fifth."

In 1941 he enlisted in the Navy after the Army rejected him. He was assigned to naval radio training school at the University of Wisconsin in Madison, where he handled public relations and edited all on his own a school paper, *The Badger Navy News*. "Buzz" Gardner served his last two years of duty on a destroyer escort, the *U. S. S. Pope*, looking for German submarines in the North Atlantic. They made him yeoman, who was responsible for the ship's log, because he knew how to type. He continued to write throughout his military service and contributed stories and articles to *The University of Kansas Review* and other magazines. Night watch gave him plenty of time to think up "crazy plots" for more stories. Though he never saw any military

* Desperate Desmond was the titular villain of a humorous comic strip burlesquing melodramas, created in 1910 by Harry Hershfield, "The Jewish Will Rogers." – Ed.

action, he learned on board how to swear like a sailor. He was never any good at it; I never knew him to utter a single four-letter word.

After the war, under the GI Bill, he returned to the University of Chicago but lasted only a year in the graduate program. He had no desire to become a teacher, so he never pursued a master's degree. He wanted to be a writer. The only course he ever took was on the philosophy of science, taught by the great Viennese philosopher and atheist Rudolph Carnap. He was of "the logical positive school," and taught the Navy veteran that "all metaphysical statements are totally meaningless in the cognitive sense." While admiring Carnap's dialectic, Martin was still able to retain some religious convictions as a "fideist" or "logical positivist." He attended a debate in which Carnap and Bertrand Russell argued whether their wives existed or not. At least that was what Martin *thought* they were saying: he never really understood the point of the discussion. Carnap's instruction so impressed Martin that years later he asked if Carnap would have his classes taped; Martin rewrote them as the book *Philosophical Foundations of Physics* (1966).

He continued to contribute to small journals for little or no pay. "Magician" was how he was identified when *School Science and Mathematics* published his article "Magic for Elementary School Science Class" in 1944. He thought he actually might be able to make a living as a freelance writer when he sold a comic short story, "The Horse on the Escalator," to *Esquire* in 1946. Reaction to the story was favorable, including a letter to the editor Martin himself penned under a pseudonym, which the magazine published. *Esquire* bought his stories regularly for about a year, and his science fiction tale "Thang" was picked for *The Science Fiction Stories and Novels* (1949). His article on writing "Humorous Science Fiction" appeared in *The Writer* in 1949, as well as stories in *Argosy* and *London Mystery Magazine*, and articles in *Philosophy of Science*, *Ethics*, and *The Journal of Philosophy* the next year. Unfortunately, the sales to *Esquire* he so depended on dried up when the magazine moved from Chicago to New York and changed editors. Martin started a novel but quickly realized it was unpublishable so he put it aside.

He knew it was time to move to New York, "where all the action is for writers," and so he did. After he published an article on pseudoscientists, "The Hermit Scientist" in *The Antioch Review* in December 1950, a friend secured a contract from Putnam's for Martin to write his first book, *In the Name of Science* (1952). Here he ridiculed the Flying Saucer craze of the late 1940s and early 1950s, L. Ron Hubbard's Dianetics, Austrian psychiatrist Wilhelm Reich and his orgone box for increased sexuality, astrophysicist Dr. Immanuel Velikovsky's claim that the Earth stood still for Joshua, Soviet physicist Trofim D. Lysenko's disastrous application of Marxism to genetics, Kenneth Roberts' revival of water discovery by divining rod, "the greatest of modern opponents of evolution" George McCready Price, and Gayelord Hauser's absurd nonscientific promotion of the miraculous nutritional value of skim milk, yogurt, brewer's yeast, wheat germ, and blackstrap molasses. Many of these hilarious cases of pseudoscience were first reported in reputable and often scientific journals that Martin scrutinized. (Back in high school, he himself had fallen for a report about an antigravity machine in *Science and Invention*.) The writer was flooded with letters from irate believers, particularly Reichians who swore by the treatment. They were unaware that when Martin had submitted that chapter to Reich himself, he approved it.

The book did so poorly that Putnam's immediately remaindered it. When Dover issued an expanded edition as *Fads and Fallacies in the Name of Science* in 1956 to cover the recent Bridey Murphy mania and other hoaxes, it became a bestseller. It has never been out of print since. Martin turned on the radio late one night to hear, "Martin Gardner is a liar." Almost nightly for nearly a year, Long John Nebel, an early champion of UFOs, had some crank or other on his radio show to attack the book; all the attention greatly boosted sales. *The Village Voice* sent copies to new subscribers, many of whom wrote nasty letters to the paper and canceled their subscriptions. What is most remarkable is that so many of these pseudosciences Martin debunked still flourish nearly sixty years later. The tragedy is that so many people still believe these quacks, cultists, and opportunists. "I think an important thing to realize," he explained, "is that science is on a continuum, with good science—science everybody agrees is good science—over at one end and at the other end crank science that everybody believes is crank." Martin believed that pseudoscience led to "the steady dumbing of our nation" and social disaster when introduced by politicians as public policy. "Professional astrologers now outnumber astronomers," he complained to Kendrick Frazier in 1998. "For Pete's sake, a president of the United States and his first lady were astrology buffs!" He also believed religion should play no part in politics. He agreed with Voltaire, "Men will commit atrocities as long as they believe absurdities."

Among the people Martin thanked in *In the Name Science* was Charlotte Greenwald for proofing and revising the manuscript. The two had met on a blind date through a conjuring friend. Martin's niece Dorrie told me that when he first saw Charlotte, he knew she was the one he was going to marry. He was immediately attracted to her great smile, lovely green eyes, and wonderful smell. Of course Martin had had other girlfriends before, but he was not really in love with any of them. At 5′8″ and never weighing more than 130 pounds, he never thought he was particularly attractive to women. She was from the Bronx and Jewish, but that did not matter. He offered to attend a reform synagogue for her and she was willing to go to the Methodist Church for him, so they agreed to avoid any specific religious affiliation. They wed in 1952 and soon had two sons, James and Thomas.

Once married, Martin could no longer rely on sporadic magazine sales, and sought steady employment. A friend, who was working at the Parents' Institute, readily hired him to run the monthly activity page of a new children's magazine, *Humpty Dumpty*, a title suggested by the editor's wife. Every month Martin wrote a story about Humpty Dumpty, Jr., and a poem of moral advice from the father to his son, as well as all the activity pages based on the old *John Martin's Book* he read as a child. These involved the art of "damaging the page" by coloring, folding, piercing, cutting, and pasting. "Librarians hate it," he admitted. Shari Lewis, the puppeteer, liked one of his puzzles so much she demonstrated it on her popular Saturday morning children's program. For several years, Martin was responsible for nearly half of each issue of *Humpty Dumpty*. One great advantage of the job was that he could work at home. He also contributed to *Children's Digest* and launched the short-lived *Piggity's Animal Stories* and *Polly Pigtails*, which eventually became *Calling All Girls* and later *Young Miss*. He was Polly.

Although he contributed a piece, "Logic Machines," to *Scientific American* as early as 1952, Martin's first article on math in the magazine appeared in December 1956. The response to the piece was so encouraging that the publisher, Gerard Piel, suggested to him a monthly "Mathematical Games" column starting in the next issue. Martin jumped at the chance for the steady paycheck and "half-bluffed" Piel by saying he had enough material for a monthly column. As he did not own any books on recreational math at the time, he ran out and bought all he could find. He learned quickly on the job. The earliest columns were a bit elemental, but soon professional mathematicians from all over the world began sending him ideas he could use in the magazine. The column became increasingly more sophisticated and complex as Martin was gaining more experience himself. He also picked up ideas from a group of magicians who met every Saturday at restaurant in lower Manhattan. "I'm defining [recreational mathematics] in the very broad sense to include anything that has a spirit of play about it," he explained. The publisher gave a free rein to write on anything that amused him at the time. "Mathematical Games" soon became the journal's most popular feature. He wisely retained reprint rights to all his contributions, and they eventually filled fifteen books.

After a few months, writing the column along with his regular chores at *Humpy Dumpty* proved to be just too much, so he left the children's magazine to concentrate on *Scientific American*. This regular job gave him the time, opportunity, and freedom to write books only when he felt like it. Although he was technically merely a freelancer with *Scientific American*, the magazine treated him as one of the staff and provided him with medical, retirement, and other benefits. Elwyn R. Berlekamp, John H. Conway, and Richard K. Guy declared in *Winning Ways* (1982) that Martin "brought more mathematics to more millions than anyone else." It is no surprise that many of his readers became mathematicians themselves. As *The Economist* acknowledged, "Mr. Gardner made mathematicians out of children and children out of mathematicians." He made geeks of us all.

Like Edgar Allan Poe and Mark Twain before him, Martin loved a good intellectual hoax. Perhaps his most talked-about column was his 1975 April Fool's joke in which he pointed out, among other things, the "flaws" in Einstein's Theory of Relativity and that Leonardo Da Vinci invented the flush toilet, complete with a diagram supposedly from one of his notebooks. Oddly, some scientists were duped by some of the examples but not all: one mathematician thought Martin should be expelled from the American Mathematical Society for not revealing in the next issue that it was all a joke. His tongue-in-cheek essay "The Irrelevance of Conan Doyle" argued that the man who believed in the Cottingley Fairies could not possibly have written the Sherlock Holmes stories. As "George Groth," a character in the semi-autobiographical *The Flight of Peter Fromm* (1973) and a pseudonym he had used as far back as 1936, Martin published a scathing review of his own confessional *Whys of a Philosophical Scrivener* in the otherwise staid and stuffy *New York Review of Books* of December 8, 1983. As Mr. Groth pointed out so eloquently in the review, Martin Gardner was not afraid to defend "a point of view so anachronistic, so out of step with current fashion, that were it not for a plethora of contemporary quotations and citations, his book could almost have been written at the time of Kant, a thinker the author apparently admires."This mock notice was so vicious that some of his friends decided not to buy the book!

In 1960, the column introduced the wild escapades of Martin's fictional alter ego, the mad numerologist and whacky con artist Dr. Irving Joshua Matrix, the supposed reincarnation of Pythagoras who, with his lovely Eurasian daughter, Iva Toshyori (with whom Martin was supposedly romantically interested), could come up with the symbolic meaning of any "randomly arranged group of numbers." In January 1961, the old humbug predicted that John F. Kennedy would win the presidential election because every twentieth-century president except Eisenhower had a double letter in his first or last name; Eisenhower squeaks by with his initials, "D. D." (One reporter, who spoke with Dr. Matrix on the phone, thought he "sounded like Martin Gardner with a clothespin on his nose.") Dr. Matrix applied math to romance as much as to politics. "Just now I'm worried about Lynda Byrd Johnson," he said in late 1967. "She should have married George Hamilton, you know. The letters in the name Hamilton add up to 92 against 95 for Johnson, and that's a beautifully close match. But now it seems that Lynda Byrd has decided on a fellow named Robb, who only adds up to 37. But let us hope such obstacles can be surmounted by love and understanding." Not everyone got the joke. In June 1974, Martin inadvertently sent gullible readers off to Pyramid Lake, Nevada in search of Dr. Matrix, who claimed his miraculous plastic pyramids were guaranteed not only to sharpen razor blades and preserve meat but also "cure all bodily ills, raise my intelligence, strengthen my psi powers, and build up my sexual potential."

Dr. Matrix was not the only humbug Martin enjoyed. He never lost his love for Oz the Great and Terrible. Martin unashamedly believed L. Frank

Baum to be "the greatest writer of children's fiction yet to be produced by America, and one of the greatest writers of children's fantasy in the history of world literature." He just did not understand the general critical neglect of the creator of Oz. No one taught Baum's books in school, and most libraries in the country banned them. "When a new book by an American about children's fiction comes along," he confessed, "I have a simple test for deciding whether it is worth reading. If it fails to mention L. Frank Baum, the author is an ignoramus." As Baum's centennial was approaching in 1956, Martin wrote the first important article on Oz to appear in a national magazine in many years. He looked up anyone remotely connected to the famous children's book author, such as Ralph Fletcher Seymour, who hand-lettered Baum's *Father Goose, His Book* back in 1899, and the daughter of Paul Tietjens, the composer of the 1902 *Wizard of Oz* musical extravaganza. Although he had never read her Oz Books, he even wrote Ruth Plumly Thompson in January 1955, asking if he might speak with her about Baum in Philadelphia on his way back to New York from Washington, D. C. She was willing to get together, but admitted that she "never met any of Baum's family and all I know of him was told me by Johnny Neill when he illustrated my books."

Martin also contacted Jack Snow, "a cheerful, rotund little man who was probably the ranking expert on the fanciful land of Oz." Snow had already written two Oz Books, *The Magical Mimics in Oz* (1946) and *The Shaggy Man of Oz* (1949), as well as compiled the encyclopedic *Who's Who in Oz* (1954). He pulled together the finest private collection of Baumiana before selling most of it off to pay his rising medical bills. For years Snow had been gathering material for a biography of L. Frank Baum he would never write and happily passed some information on to Martin. Snow later wrote another Oz fan, "I helped Martin write the article, supplying most of the information." He probably provided most of the errors too.

The publication of Martin Gardner's two-part article "The Royal Historian of Oz" in the *Magazine of Fantasy and Science Fiction* in January and February 1955, along with the recent re-release of MGM's famous 1939 movie of *The Wizard of Oz* and the centennial exhibition of Baum's work at Columbia University, did more than anything else to fire a revival of critical interest in Oz. (Martin recalled going through the Columbia retrospective with Jack Snow as he pointed out treasure after treasure he once owned.)

One important person hated Martin's article. Frank Joslyn Baum, the writer's oldest and wayward son, was so furious by what Seymour and Eunice Tietjens, the composer's widow, were quoted as saying about his father in "The Royal Historian of Oz" that he threatened to sue Martin and the magazine. Then in his next letter he invited Martin to co-write a biography of his father with him. He had offered the job to Snow, who let it slip. When Martin looked at what Frank J. had already written, he recognized immediately how difficult it would be to work with him, so he turned him down. Martin did not like the nasty things the son had written about his own mother. Frank J. Baum eventually contacted Russell P. MacFall, Night Editor of the *Chicago Tribune*, and they co-wrote the first (and highly inaccurate) Baum biography, *To Please A Child* (1961).

While recommending Snow's *Who's Who in Oz* as "a *must* for every reader who enjoyed Martin Gardner's recent *F&SF* articles on Oz and its creator," the editor Anthony Boucher wondered, in the April 1955 issue, "why these deathless chronicles have not attracted a group of enthusiastic amateur scholars similar to the Baker Street Irregulars. If any of you are interested in an informal organization for the purpose of relishing Oz as the BSI do the canon of Sherlock Holmes, drop me a note and I'll see that it reaches Mr. Snow, who should be the logical focus of such a group." Martin, William Baring Gould of *Life* magazine, Fred Dannay (half of the "Ellery Queen" writing team), and others were eager to form such an organization. But Snow dreaded the thought of having to correspond with every Oz nut in the country to take up Boucher's challenge. Suffering from jaundice and dropsy from years of alcohol abuse, Snow died on July 13, 1956. Martin, serving as Snow's sole executor, placed his obituary in *The New York Times* the following day and disposed of his personal papers and what was left of his once legendary Baum collection. Many Oz collectors benefited handsomely from Martin's generosity. He kept only a few things for himself, some books, papers, and a rare poster of L. Frank Baum that was issued by George M. Hill upon the publication of *The Wonderful Wizard of Oz* in 1900.

Baum fans, including Martin, had long talked about taking up Boucher's challenge and forming an Oz society like The Baker Street Irregulars, but it was not until 1956 that the thirteen-year-old boy collector Justin G. Schiller took the initiative with schoolteacher Fred M. Meyer to found The International Wizard of Oz Club. They contacted everyone who had corresponded with Snow and launched the organization with only fourteen charter members. To Martin's great amusement, he was listed as Chairman of the Board on the masthead of the little mimeographed club journal, *The Baum Bugle*. Martin attended some of the meetings but eventually lost interest. The club was always collector driven and that was not Martin's idea of an Oz society. Possessions never meant much to him beyond their immediate practical use.

In 1957, when *The Wizard of Oz* went into public domain, Michigan State University Press published Martin's revised "The Royal Historian of Oz" in the first scholarly edition of the book, titled *The Wizard of Oz and Who He Was*. The endnotes Martin supplied to Baum's American fairy tale made it the first "annotated" version of the story. It also contained an appreciation of Baum's work by Pulitzer Prize–winning historian Russel B. Nye of the school's English Department. Although Oz was not taught in college courses at the time, some academics actually embraced the book. "In view of the absurdly snobbish attitude which the half-educated take toward L. Frank Baum," wrote Dr. Edward Wagenknecht of Boston University in his review in the *Chicago Tribune*, "it delights my soul that this book should be published by a university press." The only previous scholarly monograph on Oz was Wagenknecht's own *Utopia Americana*, published by the University of Washington in 1929.

The Wizard of Oz and Who He Was came out just in time for an ugly literary debate. When the director of the Detroit Public Library, Ray Ulveling, denounced *The Wizard of Oz* at a librarian meeting in 1957, Martin, Nye and the university press all came to Baum's defense. Ulveling seemed smug about keeping *The Wizard of Oz* off the children's library shelves: Baum's stories were "of no value" with "nothing uplifting or elevating" about them. Martin was so outraged that he became one of the most vocal opponents to the then almost universal banning of the Oz Books from the nation's libraries. "To this day," he reported in "Why Librarian Dislike Oz" in *American Book Collector* (December 1962), "there are thousands of children's librarians around the nation—kindly, bespectacled, gray-minded ladies—who will tell you with pride that they do not permit a single Oz book to sully their shelves." He suggested the problem lay not with Baum but in the "prosaic, matter-of-fact mind" of the typical librarian: "An individual with a soaring imagination is not likely to be happy shuffling file cards." Martin also provided lively introductions to Dover's reprints of many of Baum's children's stories, some long out of print. It was largely through Martin's persistence that the Oz Books were eventually made available in children's libraries across the country.

It took Martin five years to get a publisher interested in his pet project, *The Annotated Alice* (1960). He at first proposed that Bertrand Russell write the book (Russell even looked like Tenniel's Hatter); when he declined, an editor at Dial, Clarkson N. Potter, suggested Martin write it himself. No one else was more suited for the task of annotating *Alice*. Martin Gardener in a sense was a modern Lewis Carroll. He shared the Oxford don's ragbag intellectual curiosity. Both men were fascinated with philosophy, mathematics, logic, literature, and puzzles. He shocked old-timer Carroll fans by concluding that "the last level of metaphor in the Alice books" was "life, viewed rationally and without illusion, appears to be a nonsense tale told by an idiot mathematician."

The Annotated Alice was *not* written for children. "I have not yet formed an opinion of how well the children of Carroll's day liked his *Alice* books," he once wrote the penultimate Oz fan, Fred M. Meyer. "(It is hard to know, because children's books are usually bought by parents, and I suspect that even then most of the sales of *Alice* were due to the fact that adults were charmed by the story, and that there was very little else available for children that didn't involve pious moralizing.) I do know that I have talked to a great many people who are very fond of Alice, but tell me that they hated the books as children. This was my very own experience. I did not appreciate them until I reread them in my twenties. On the other hand, I have yet to meet someone who tells me that he read the Oz Books as a child, but didn't like them." Martin believed that, unlike the vast fan base for *The Wizard of Oz*, the audience for *Alice* was severely limited. "I can't imagine a young child, say under seventeen, getting much of anything out of the two *Alice* books," he admitted to physics professor Donald Simenek in 2007. He maintained that anyone "who is bored by magic tricks, dislikes puzzles, and has never played a chess game" could not possibly "ever get to the center of Carroll's mind or fully appreciate all his writings." Nevertheless, despite the author's intention, *The Annotated Alice* for decades introduced countless boys and girls to the wonders of Wonderland.

Only one publisher would have taken a chance on such a peculiar work. Clark Potter shared Martin's fascination for the obscure and the amusing. (He was later forced out of the company he created.) He issued an ambitious, sumptuously produced volume with all the original John Tenniel illustrations that sold in 1960 for the then astonishing price of $10.00. Martin followed *Alice* with the less successful *The Annotated Snark* (1962), as well as *The Annotated Ancient Mariner* (1965), *The Annotated Casey at the Bat* (1967), *The Annotated Innocence of Father Brown* (1987), and *The Annotated Night Before Christmas* (1991). Martin encouraged other writers like William S. Baring-Gould, who produced, with his wife Ceil, *The Annotated Mother Goose* (1962) and the even more successful two-volume *The Annotated Sherlock Holmes* (1967). He also convinced Potter to sign up his great friend Isaac Asimov for *The Annotated Gulliver's Travels* (1980). That nearly killed the series. The manuscript arrived in such a dreadful state that it is no surprise Asimov was able to put out as many books as he did: they were not so much written as rewritten by his editors. He never did his own picture research, so the publishers hired me to do it for him. The Annotated Books remained Clarkson N. Potter's best known titles until the company unleashed the ubiquitous Martha Stewart.

Martin cast a wide net of enthusiasts, and his best friendships were usually epistolary. For example, he met the great science fiction writer Arthur C. Clarke only once for lunch and yet they kept up a lively correspondence for years. He was particularly fond of Clarke's observation, "I sometimes think that the universe is a machine designed for the perpetual astonishment of astronomers." Shy by nature, Martin preferred one-on-one relationships. He hated crowds, giving speeches, and going to parties. He did belong to the Trap Door Spiders, a secret stag club of mostly science fiction writers (including Isaac Asimov, L. Sprague de Camp, Lester del Rey, Lin Carter, Stefan Kanfer, and James "The Amazing" Randi) who met for dinner on one Friday a month in New York. "Nothing pleases me more," he told Kendrick Frazier, "than to be alone in a room, reading a book or hitting typewriter keys." He received thousands of letters from all over the globe. The earliest readers of "Mathematical Games" were high school students, but soon serious and not so serious mathematicians generously sent Martin more games and solutions than he could possibly use in the column. He once admitted that he never really understood all that they wrote him, but saved their letters anyway. He rarely failed to answer a total stranger no matter how baffling the letter might be.

Among Martin's literary friends was William Lindsay Gresham, best known for his novel *Nightmare Alley* (1946). Both men shared a deep love for the Oz Books and Martin invited him to join the club. An early devotee and later denouncer of Scientology, Gresham exposed the scams of sideshow spiritualists and wrote *Houdini: His Life and Art* (1976) with Randi. Gresham was also an alcoholic and a womanizer. His wife, the poet Joy Davidman, informed him one day that she was leaving, taking their two sons and heading to London, where she was going to marry the English writer C. S. Lewis, whose work she greatly admired. He thought she was crazy. She did just that! Gresham quipped that Lewis would now have to change the title of his famous book about his religious conversion from *Surprised by Joy* to *Overwhelmed by Joy*!

After Joy died from bone cancer, Gresham went to England to talk to Lewis about his sons. He had a copy of *The Annotated Alice* signed for Lewis and asked Martin if there was anything he wanted him to ask the author of the Narnia Chronicles. Yes, Martin was curious if the English writer had ever read the Oz Books. Unfortunately, Lewis told Gresham that he never did. Gresham confessed to Martin that when his marriage broke up, he contemplated taking his own life until he came across a copy of *The Scarecrow of Oz* while staying at a friend's house. Martin suggested he write about the incident for *The Baum Bugle*. "It came like a triumphal chorus, an Ode to Joy," Gresham recalled, reading *The Scarecrow of Oz* after all those years, "that Oz *is* True—every golden word of it." Gresham eventually did commit suicide when he was going blind and developed throat cancer; Martin told me he did not want to burden anyone.

Martin finally concluded that all the effort he spent on debunking pseudoscience was "a waste of time." He far more enjoyed writing books on Carnap, science, math, and other subjects that really interested him, such as Einstein's theories in *Relativity for the Million* (1962) and *The Ambidextrous Universe* (1964), his study of dichotomy. One of the biggest fans of *The Ambidextrous Universe* was the Surrealist painter Salvador Dalí, who wanted to meet Martin when he and his entourage came to New York. They shared a passion for mathematics and Alice. Dalí seemed to Martin "perfectly normal," but he brought with him an aspiring actress who called herself

Ultra Violet and claimed she was Venusian. Dalí thought Martin should write her biography, but he gracefully declined. She later ran off to join Andy Warhol's Factory.

REMEMBERING THE MASTER ANNOTATOR

The Ambidextrous Universe was required reading in one of two science courses I took at liberal-arts Bard College in Annandale-on-Hudson, New York. Of course I was already familiar with Martin Gardner's work. At ten, I had fallen in love with *The Annotated Alice* and knew his name through The International Wizard of Oz Club that I joined about the same time. Somehow I got his address when I was in high school and wrote to urge him to do *The Annotated Wizard of Oz*. He replied that he did not think he was qualified. He felt the "annotator" had to be familiar with the entire Oz series and he had only read the books through *The Royal Book of Oz*, believing it was by Baum, and the two Snow titles. He had suggested to Fred Meyer that he try it, but he gave up when he thought that, unlike the *Alice* books, there just was not enough in *The Wizard of Oz* requiring annotation.

On completing my sophomore year at Bard, I did not have a summer job, so I decided I would have to write *The Annotated Wizard of Oz* myself. I had no idea where I was heading. The English Department informed me that spring that I had no business being there. Undaunted and rather naively, I cobbled together a proposal and sent it to Clarkson N. Potter cold. He took my unsolicited manuscript out of "the slush pile," read it and liked it, and sent it to Martin for his opinion. He told Potter to sign me up. Within a month I had my first book contract, with a check for the princely advance of $1,500 against royalties on signing. I was twenty years old at the time.

No one questioned my age or ability to prepare *The Annotated Wizard of Oz* until I wrote the Children's Book Editor of *The New York Times Book Review* to see if he might be interested in an article on my project. A few days later I found a note in my mailbox at Bard, requesting me to call George A. Woods collect at the *Times*. When the editor answered the phone, he said, "Thank you for calling me back, Professor Hearn." I immediately corrected him: I was not a professor. "Oh, you're not?" "No, I'm a junior at Bard."There was dead silence on the other end. "Oh," he finally said. Smelling a story here, Woods hired Martin to write a lead article on the entire Oz phenomenon.

On May 2, 1971, the front page of *The New York Times Book Review* carried "We're Off to See the Wizard," which began, "A young student of English literature at Bard College, Michael Patrick Hearn, is busy on a curious project. He is writing scholarly footnotes for a volume to be published next year in Clarkson Potter's well-known series of annotated classics. The book? *The Wonderful Wizard of Oz*!" Martin published the address of The International Wizard of Oz Club in his article, and membership swelled to over 1,000 for the first time in its history. The volume of mail it engendered so swamped Secretary Fred Meyer that he almost quit!

While I was working on the manuscript, Martin invited me to visit him at his lovely home at the mathematically appropriate 10 Euclid Avenue in Hastings-on-Hudson, New York. I was able to come on a Sunday afternoon, and Martin met me at the train station and drove me back to his house. We had so many things to talk about we did not know where to begin or when to stop. He wanted to learn more about me. He sat up straight in a stiff-backed chair in the sunny living room with his arms folded across his chest, caterpillar-style. He could stay in this position for hours without even a twitch, conversing without end in his sweet, soft, gentle, high-pitched monotone. His laugh was barely a whisper. Martin explained to me that Potter's Annotated Books from the first were designed as remainder books. (Crown Publishers, who bought Clarkson N. Potter Inc., dropped the price of $10 to $3.95 for the cheaply produced editions of *The Annotated Alice*.) He warned me that I should not expect much of a royalty check from the original trade edition. It was only when the publishers issued the more affordable reprint of *Alice* that he began to earn any real money as it went through printing after printing. Readers were always sending him new material and the format of the Annotated Books was so flexible that he easily inserted new notes every time it went back to press, and he encouraged me to do the same. Martin also mentioned that, as a sign of his deep affection for Baum, he always tried to mention Oz somewhere in every one of his books. In going through his works recently in preparation for this essay, I suddenly noticed that he referred to me too in many of them.

Stieg Larsson describes in the second volume of the bestselling "Millennium Trilogy," *The Girl Who Played With Fire* (2006), how his punk heroine "advanced through Archimedes, Newton, Martin Gardner, and a dozen other classical mathematicians." Martin would have found his being linked with the greats both highly amusing and inaccurate. He would not have considered it a compliment. One of the first things he ever told me about himself was, "I'm really a fraud. I'm not a mathematician. I never had any more math than high school." His having never earned a doctorate irked some academics. "I couldn't solve a calculus problem if my life depended on it," he once admitted. He never produced a column about it either. He was a journalist who *wrote* about mathematics. "The big secret of my success as a columnist was that I didn't know much about math," he told *The New York Times* in 2009. "I had to struggle to get everything clear before I wrote a column, so that meant I could write it in a way that people could understand." As for any mathematical games, he admitted that he may have invented only a few entirely on his own. He explained to Rucker, "I'm like a person who loves music and enjoys listening to it but who doesn't compose or even play very well." He just wrote about it clearly and beautifully and with a wry sense of humor.

When I asked him a question he could not answer off the top of his head that afternoon in Hastings-on-Hudson, he suggested we go up to his study. It was unusually neat and tidy for a writer's workplace. His extensive library encompassed often rare and obscure books and periodicals on mathematics, puzzles, magic, physics, literature, philosophy, religion, and countless other subjects. (To Charlotte's chagrin, he never shopped for clothes, only books. He used to write me whenever he found some new treasure.) He generously loaned me some Baum titles I could not locate, most being copies from his own childhood and extensively annotated by him in the margins in pencil. He also kindly gave me permission to quote anything from *The Wizard of Oz and Who He Was* I chose. The breadth of his reading and knowledge was, well, breathtaking. I learned of quacks and quarks and cracks in the universe from him. He was practically the only science writer I could ever comprehend.

He kept the most incredibly deep card files of anyone I have ever known. As I asked him one obscure question after another, he went to his cabinet and pulled out one three-by-five card after another containing some perfect and perfectly arcane information. Like a housewife collecting future recipes, Martin never hesitated to snip from a book or periodical some passage that caught his fancy and paste it to one of his cards. He considered himself no more than "a scissors and rubber-cement man." This was a habit he developed early on and had accumulated thousands by the time he returned to Chicago during the Great Depression. "Those cards, filling some twenty-five [ladies'] shoe boxes, are now his most precious, and almost only possession," reported *The Pulse* in 1940. "The card entries run from prostitutes to Plautus—which is not too far—and from Plautus to police museums." A rare book dealer was horrified when he realized that some of them contained passages cut out of a valuable first edition of *The Great Gatsby*! Martin's manuscripts were often a regular collage of typed commentary and bits from other publications.

It was getting late and Charlotte invited me to stay for dinner. Then I dropped the "V" word—vegetarian. She was flummoxed at first but then instructed Martin to order a pizza. Apparently "calling out" was an uncommon occurrence in the Gardner household. Martin went to pick up the order and returned with it under his arm. "Martin!" Charlotte sighed. "You don't carry a pizza under your arm!" He looked at her and he opened the carton and saw that the cheese had all run down the side of the pie. "Oh," he said calmly, "you're right." Charlotte took it from him and rearranged the topping and it was fine. During dinner, Charlotte said something to their son Tom he did not like and he shot back and they began to argue. It was a typical mother and teenager spat, but Charlotte suddenly burst into tears at the dinner table. I understand now that it was just not like her to cry over anything. Perhaps the awkwardness of my being there to witness this little domestic quarrel caused the breakdown. What was remarkable was that Martin did not strike or chastise or punish his son in any way. He merely said, "Oh, Tom, now you've made your mother cry." There was no rebuke in his voice—just deep disappointment. Fortunately the crisis quickly passed and Martin and I continued our discussion after dinner back in the living room. We kept losing track of time and missing "the next train" until Martin had to rush to get me to the very last one out that night.

I saw Martin later that summer on a sweltering Saturday in August at the Munchkin Convention sponsored by The Wizard of Oz Club in a little church in New Jersey. I got a ride down with Justin Schiller and his partner, Ray Wapner, and Martin brought his sister Judy, also a longtime Oz fan, and her teenage daughter Dorrie. I believe it was my first Munchkin Convention. It may well have been Martin's first, too—and his last. The meeting was held in the church's basement, which was not equipped with air conditioning, and I noticed how uncomfortable Martin was throughout the proceedings. There was the customary Oz Quiz in the morning. Of course Martin aced the first part devoted to Baum's stories, but he was stumped by the second that dealt with the Ruth Plumly Thompson Oz Books. He had never read them, so I helped him out by whispering the answers, which he wrote down.

None of us was impressed with the cafeteria-style lunch the church provided so Martin, Judy, Dorrie, and I went off to find a nice coffee shop in the area. Judy could not stop singing the praises of her wonderful brother. Martin hardly said a word. Did I know that the English writer John Fowles had quoted *The Ambidextrous Universe* in his recent bestseller, *The French Lieutenant's Woman* (1969)? Vladimir Nabokov, too, mentioned her brother in his latest novel, *Ada, or Ardor* (1969). Martin explained that they were both Lewis Carroll fans. When Martin quoted in *The Ambidextrous Universe* two lines from the epic poem that made up the bulk of the Russian's earlier novel *Pale Fire* (1962), he credited them to the fictional poet "John Shade" rather than to Nabokov. The novelist was so amused that when he quoted the same lines by "John Shade" in *Ada*, he credited them to *The Ambidextrous Universe* by "an invented philosopher" named "Martin Gardiner" [*sic*]. Martin suspected that Nabokov's last novel, *Look at the Harlequins!* (1974), owed much to *The Ambidextrous Universe*.

Although never interested in the collectable aspects of Oz, Martin decided to tough it out and returned for the auction that made up the afternoon session of the Munchkin Convention. He had donated a copy of the rare *The Woggle Bug Book* (1905) and was curious how it might do. I sat next to him as the bidding, much to Martin's surprise and discomfort, went higher and higher and higher. He turned to me and asked if it was really worth that much money. "Yes," I whispered, "I don't even have a copy!" "Oh, Mike," he said crushed, "Had I known, I would have given it to you!"

Thanks to Martin, I was able to complete the manuscript of *The Annotated Wizard of Oz* within a year (probably the only time I ever met a publisher's deadline) and hitchhiked from Bard down to New York City to personally deliver it to Potter in his office. He graciously greeted me and then excused himself. I heard him shout to someone in another room, "How old *is* he?" Potter published the manuscript almost exactly as I wrote it. The only objection my editor, Jane West, made was to my feminist interpretation of Oz. She told me Women's Liberation was a passing fad and did not want to date the book. I strongly disagreed, but slightly toned it down. The company wisely decided not to publish an author's picture on the jacket in case a reviewer assumed this kid was unsuited to write such a scholarly book. It is remarkable how many people still believe Martin wrote *The Annotated Wizard of Oz* and I *The Annotated Alice*. Did they think he was my ghostwriter? Whenever anyone mentioned how much he or she enjoyed "your book, *The Annotated Alice*," I replied, "I know Martin Gardner will be very happy to hear that."

Few people know that Martin was indirectly responsible for the "super soul musical" *The Wiz*. I am not sure even Martin knew! Radio DJ Ken Harper wanted to break into Broadway production, and Ken told me later that reading Martin's piece on Oz in *The New York Times Book Review* gave him the idea for a new, all-Black musical based on *The Wizard of Oz*. The article convinced him to go right back to Baum's book rather than to merely revamp the famous Judy Garland picture. Martin told me how much he enjoyed the new show's score when composer-lyricist Charlie Smalls premiered it at the book party for *The Annotated Wizard of Oz*. Had we only invested in that show! It would have made both of us wealthy. Ken wanted Martin to write the script and critic Gene Shalit to host a television documentary on L. Frank Baum (Ken was sure Baum was gay!), but nothing came of the project. Ken died of AIDS before he could produce his second musical, *Bamboo*.

While waiting for *The Annotated Wizard of Oz* to come out (Crown delayed it over a year) and working on my second book, *The Annotated Christmas Carol* (1976), I applied for my first job after college. I had read in *The New York Times* that a new children's magazine called *Cricket* was looking for an editor. I asked Martin if he would be willing to write a recommendation for me and he graciously complied. I did not know when I applied for the position that the Caruses, who owned the magazine, were great fans of his *Scientific American* column. I learned later that I got the job as *Cricket*'s New York Editor primarily because I knew Martin Gardner.

The Caruses were eager that Martin introduce a monthly puzzle page for young readers in *Cricket* so I approached him with the idea and he ran with it.

He had always wanted to do his own feature based on "Peter Puzzlemaker" in *John Martin's Book* that he so loved as a child. He at that time was trying to collect the entire run of the long forgotten children's magazine. "The use of old puzzles will be inevitable (there just aren't that many brand new ones)," he explained, "but every effort would be made to dress up the oldies in novel ways, and to use as many new ones as possible." He dashed off a detailed six-page proposal for "The Funny Puzzles of Phineas Phun" with a single sample, which was turned over to *Cricket*'s Art Director, Trina Schart Hyman, for pictures. She hated it! She thought the concept and characters were far too old fashioned and "uncool" so she dawdled with the drawings until both Martin and the Caruses lost interest in the feature. It was a shame. A Martin Gardner puzzle page would have greatly enlivened *Cricket*. Perhaps it was just as well: the sample puzzle involved matches. (I suggested he change them to toothpicks.) Also, Martin admitted, "A magazine like *Cricket* would never think of having to cut the page."

It was through Martin that I became a charter member of the Lewis Carroll Society of North America. He invited me to join him and a group of especially enthusiastic Carrollians to form an American branch of the Lewis Carroll Society of the U.K. over dinner in Princeton. Martin treated me to my first Manhattan that evening—and my last. I was flying the rest of the night. A number of the Society's early members, including Justin Schiller, Martin, and myself, were recruited from The International Wizard of Oz Club. Although Martin submitted an article from time to time to both clubs' publications, he stopped going to the meetings. He felt the general academic discussion of Carroll's and Baum's masterpieces tended to veer towards the dry side. Martin could not have cared less if there was broken type in the line third from the bottom of p. 127 in the seventeenth edition of *Pirates in Oz*.

Martin Gardner did not suffer fools and frauds gladly. He made many enemies over the years, especially in academia, where many never considered him a serious scholar. He was merely a "popularizer" of math and science, they said. "Psychic hucksters" like Uri Geller, the celebrated spoon bender, and "thought" photographer Ted Serios particularly annoyed him. (He published two parodies, *Confessions of a Psychic* in 1975 and *Further*

Confessions of a Psychic in 1980, under the pseudonym "Uriah Fuller.") He, along with James Randi, Isaac Asimov, Carl Sagan, Stephen Jay Gould, and B. F. Skinner, formed the Committee for the Scientific Investigation of Claims of the Paranormal (CSICOP) and contributed a regular column, "Notes of a Fringe Watcher" to *Skeptical Inquirer*. He pretty much thought everyone involved with parapsychology was a charlatan.

Publishers, too, vexed him. He was astonished when Jane West, our editor at Clarkson N. Potter, published Michael Eddowes's peculiar *The Oswald File* (1977), in which he claimed John F. Kennedy was really killed by a Soviet agent *impersonating* Lee Harvey Oswald. "I don't think Jane was happy about my comments on the Oswald book," he wrote me. "I really thought it had been published cynically, for the money, but it turned out that Jane thinks it is a really significant book!" He called Nat Wartels, who founded Crown Publishers, a scoundrel. *The Annotated Wizard of Oz* never sold as well as *The Annotated Alice*; and Crown readily let it go out of print once the initial printing was exhausted. They did even worse with *The Annotated Christmas Carol* and *The Annotated Huckleberry Finn* (1981). Martin and I had a running competition as to who would have to wait longer for his royalty check from Crown. I won: it took me eight years to get $4,000 out of them. So much of a writing life is dependent on faith in those with whom one must work. Neither Martin nor I ever figured out exactly how many thousands of dollars in earnings Crown cheated us out of. I suspect he was the bigger loser in the long run.

By 1990, Martin had gathered such a wealth of new material on the two *Alice* books that he wanted to issue an expanded version of *The Annotated Alice*, but Crown turned him down. The publishers said the regular edition was selling well enough and refused to invest any more money in it. Fed up with that firm, Martin took *More Annotated Alice* to the parent company, Random House, who had bought Crown in 1988. Random issued in 1991 a handsome, elegant and far more elaborate volume than the cheap remaindered *Annotated Alice*. Because he did not want the new book to compete with the old, Martin decided to illustrate it with pictures by Peter Newell instead of Tenniel's classic wood engravings, and he asked me if I could tell him anything about Newell. I had grown up on my mother's copy of the Newell edition, and by chance I had recently published an article on the once famous American illustrator of *Alice* and sent it to Martin as a reference. He liked it so much he asked if he could reprint it in his book. I said he could have it for free, but he sent a check anyway. I am sure the money came out of his pocket and not the publisher's.

Martin always intended to eventually meld *The Annotated Alice* with *More Annotated Alice*, but Crown persistently refused to comply. Even when their remainder edition was actually remaindered, they would not return the rights to Martin because the paperback edition was still in print in England. It was all smoke and mirrors, and Martin finally took back both editions and submitted the expanded combined version to his favorite editor, Robert Weil at W. W. Norton. The publication of "The Definitive Edition" in 1999 revived the Annotated Books. At Martin's urging, Bob talked me into expanding *The Annotated Wizard of Oz* for the 100th Anniversary of *The Wizard of Oz* in 2000. The series has continued to grow and prosper at Norton far beyond any previous attempt by any other publisher. After the release of Tim Burton's *Alice in Wonderland* in 2010, Martin was pleasantly surprised to receive a royalty of $75,000, the largest check he had ever received in his long literary career.

What he once said of Stephen W. Hawking's prose is true of his own: "as informal and clear as his topics are profound." Martin's manner of writing was easy, precise, conversational, and never pretentious. He adhered to his hero Bertrand Russell's advice: "whenever he thought of a simpler word to substitute for a more complicated word, he would use the simpler word." Only an academic could fail to understand or appreciate him. Martin could not have cared less. F. X. Murphy, the Redemptorist, characterized Martin's manner as "a whirlwind of quotations, poetic effusions, and the paraphernalia of learned as well as pedestrian discourse." Canadian critic Hugh Kenner went so far to say in *The New York Times Book Review* that Martin's style was "plain to the point of naïveté." That is too harsh but expected: the two men clashed on the importance of Ezra Pound as a poet. Martin thought he was "a total fake." He did not care much for W. B. Yeats and William Carlos Williams; T. S. Eliot was "overrated." He had no use for "New Criticism" either.

Martin detested free verse, except that by Stephen Crane and Carl Sandburg. He liked "classical" narrative poems, ones with melody, rhyme, and meter that stuck in the mind and never left. Doggerel was good enough for him if it was good doggerel. Although he was the last to admit it, Martin was a skillful nonsense poet himself. Pick up a copy of his 1969 collection *Never Make Fun of a Turtle, My Son* (if you can find one), read any verse at random and see how lithe and lively the rhymes are. It is no surprise that some have even been anthologized. He also composed an amusing parody of "Casey at the Bat," "Casey's Son: The Story of a Terrible Downfall," under the pseudonym "Nitram Rendrag" ("Martin Gardner" spelled backwards in Looking-Glass fashion) in his *Annotated Casey at the Bat*. Under the anagram "Armand T. Ringer," he published a nimble parody of "The Village Blacksmith" about the wrestler Jesse Ventura in his *Favorite Poetic Parodies* (2001). I tried to get him to compose an original verse or two for a couple of anthologies I was editing, but he declined. He never thought of himself as a poet. Instead he gave me the names and addresses of other people he thought would write something much better for me. I took him up on his suggestions and was not disappointed.

Martin was never afraid of expressing his opinions, no matter how unchic they might be. Adam Gopnik in *The New York Times Book Review* spoke of his "old-fashioned, almost nineteenth century, Oliver Wendell Holmes kind of American mind—self-educated, opinionated, cranky, and utterly unafraid of embarrassment." He possessed "the American gift of not caring too much what people think. He likes to argue about ideas and doesn't care if he seems overeager about some of them." But he *did* care: he cared deeply. Martin strongly objected to Prof. Arnold J. Toynbee's lumping Senator Joe McCarthy with Hitler and Mussolini. "I despise McCarthy and all he stood for," Martin wrote to *The New York Times*, "but surely to equate this pipsqueak politician with Mussolini and Hitler (omitting Stalin, who sent millions of innocent people to their death by firing squad or starvation in Siberian labor camps) is to display a curious lack of balance, unbecoming to a distinguished historian." Toynbee awkwardly and unconvincingly responded that the similarity of "these three bad men" was in their "all being demagogues of genius."

Martin scoffed at that maestro of myth Joseph Campbell, challenging his dictum "follow your bliss": "How about a person . . . whose bliss is to rape little children?" He just could not abide Campbell's anti-Semitism, racism, and conservatism. He bore no mercy for prominent televangelists like Jimmy Swaggart, Oral Roberts, and Jim Bakker, especially when they fell from public grace. He had no more use for New Age nonsense as practiced by gullible celebrities than he did for flatulating fundamentalists. "Our nation is now in the midst of unprecedented enthusiasm for beliefs that medieval astrologers would have considered insane," he reported in *The Washington Post* in 1975. "The enthusiasm is more frightening than funny." The major causes for this intellectual dilemma were "the decay of religious orthodoxy, and disenchantment with science." He accused Shirley MacLaine of "teetering on the edge of solipsism, the ultimate in self-absorption," and laughed at Kaballah as hyped by Madonna ("that great philosophical scholar") and Demi Moore. He never could fathom why so many otherwise intelligent beings could be taken in by such obvious flimflams. Why did they not just use their common sense?

Of course many people, particularly his intellectual opponents (and he had many), thought Martin was no more than "a crackpot" himself. F. X. Murphy dismissed him for perpetuating "the myth that there is an inevitable war between science and religion." Whenever one spits at a scared cow, invariably someone else will spit back. Colin Wilson, the prolific British writer and the frequent object of Martin's wrath, accused the American critic of generating "an atmosphere of such intense hysteria, reminiscent of a medieval Inquisitor denouncing heresy, or Hitler fulminating against the Jews." Martin thought Wilson was no more than one of those "amiable eccentrics" who "prowl comically about the lunatic fringes of science, looking for ever more sensational wonders and scribbling ever more boring books about them for shameless publishers to feed to hungry readers as long as the boom in occultism lasts."

Martin's literary tastes were a bit eccentric. Who was the American writer he thought most worthy of the Nobel Prize for Literature? John Updike. And why? He was the one whose virtues as an author and personal philosophy most resembled Martin's: "[Updike] is one of the few living writers of fiction who understands modern science, who can deal with sex graphically without degrading it to pornography, and who is not ashamed to call himself a theist. His light and not so light verse towers over that of any living poet writing in English. His literary criticism is of the highest order." He thought much of James Joyce's *Ulysses* but little of *Finnegans Wake*. That did not prevent him from quoting from the latter whenever appropriate. (He said he usually lifted his epigraphs from a book on *Finnegans Wake*.) Martin was one of the last people on Earth who still liked Lord Dunsany, and he was able to enjoy both H. G. Wells and G. K. Chesterton (though not his Catholicism). Chesterton left Martin with "a sense of mystery in the universe" and Wells a "tremendous interest in and respect for science." He sided with the democratic socialism of Wells and shared Chesterton's deft verbal juggling of paradoxes. Martin was, as Gopnik observed, "the critical rationalist who falls in love with the fantastic products of the repressed English imagination." He particularly liked a quote from Chesterton's 1906 essay "The Dragon's Grandfather": "In the fairy tales the cosmos goes mad, but the hero does not go mad. In the modern novels the hero is mad before the book begins, and suffers from the harsh steadiness and cruel sanity of the cosmos."

Here was Martin's basic philosophy: "There are a lot of scientists who really think that there's nothing really mysterious about the universe and in just another century or so we'll solve all the problems of science," he explained on the television show *The Nature of Things*. "I belong to that group that thinks that we are very, very far from the frontier of knowing everything in science." He called himself a "mathematical realist" and a "philosophical theist" as well as a Mysterian, one of a small group of philosophers who "are convinced that the human mind, consciousness, and free will are so profound and difficult to explain that no one has the slightest idea how the brain does it." He concluded that God's attitude was probably, "I've done the best I could do to bring your world and you into existence. Now you're on your own. Let's see how you manage it."

Martin, like Nabokov, had no patience for literary psychoanalysis. Sigmund Freud was no more than "a crackpot." Martin noticed the neat similarity between "Freud" and "Fraud." He was immensely fond of Joyce's aphorism in *Finnegans Wake*: "when they were yung and easily freudened." He strongly objected to *The Interpretation of Dreams* being included in *The Great Books of the Western World*. "If one wants to search for phallic symbols, they are always easy to find," he admitted to Fred Meyer in a letter of December 22, 1960. "The Hammerheads in *The Wizard*, the horn on the Horner's forehead, the Hyppogyraf in *Tin Woodman*, and most of all, Zog in *Sea Fairies*. There is quite an astonishing scene describing the writhing of Zog's underbody beneath his robe. I mention all this to show how easy it is to play the game of 'let's look for Freudian symbols.'" Martin was among the first to question the scientific validity of "recovered memory therapy." "The greatest scandal of the century in American psychiatry," he wrote in "Notes of a Fringe-Watcher" in 1994, "is the growing mania among thousands of inept therapists, family counselors, and social workers for arousing false memories of childhood sexual abuse."

His taste in art too was generally conservative. He defended Norman Rockwell after a John Canaday attack in *The New York Times*. "Hasn't the lovable old art critic learned yet that popular, sentimental art, in all media and in all times and cultures, has its own standards and reason for existence?" he asked. "Surely there is more to say about Rockwell than just to repeat what everybody knows: namely, that his subject matter is not the same as, say, Goya's." Modern Art baffled him. He had no use for Abstract Expressionism and "the professionally Pollacked walls of wealthy businessmen." He once made an imitation drip painting and had to take it down when too many people thought it was a genuine Pollack. It amused him to reproduce one of Ad Reinhardt's pictures in a column on abstract art in *Scientific American*. "Of course, it was just a solid square of pure black," he recalled. "The publisher insisted on getting permission from the gallery to reproduce it." Pop Art equally bored him. "Who," he wondered, "would have thought it possible to find a gimmick that would produce paintings uglier than, say, those clumps of black brushstrokes by Franz Kline or more banal than those colored stripes of Mark Rothko?" He once confessed, "Chinese written words are (to me) more aesthetically pleasing than the black brushstrokes of abstraction by Franz Kline." It was the same with music: he insisted that Beethoven's work was "superior to that of John Cage or a punk rock band." He admitted he had no ear for classical music, preferring Dixieland jazz and old melodies he could play on the musical saw.

Martin lambasted Alexander Calder's 1946 doodles for Coleridge's masterpiece in *The Annotated Ancient Mariner*. He had first encountered them in an exhibition during his University of Chicago days: they baffled him then and baffled him still. "In those days I suppose the pictures had a mild shock value," he sniffed. "Today they seem merely silly, like gay imitations of Thurber cartoons." Instead Martin chose to reprint the then unfashionable Gustave Doré's more literal interpretation in his book. "The first purpose of graphic illustration for a narrative," he reminded his readers, "is graphically to illustrate the narrative." He was no more enthusiastic for modern tendencies in children's book illustration. "It is hard enough today," he admitted, "to get young ones to look at pictures that don't move, but impossible to get them to look at unmoving pictures they can't understand. If you doubt this, try showing a little girl one of Brian Wildsmith's distorted scenes for *Mother Goose* alongside Kate Greenaway's marvelous illustrations for Browning's *Pied Piper* . . . and ask her which she likes best and why." Yet he loved the visual conundrums of modern Dutch graphic artist M. C. Escher as well as the picture books of his Japanese disciple Mitsumasa Anno. Escher was little known in America (except among mathematicians) until Martin devoted a *Scientific American* column to his work. One of Martin's prize possessions was an original Escher color woodcut, "Circle Limit III" (1959), that he picked up for sixty bucks and hung in his living room. He readily pointed out that it derived from French mathematician Jules Henri Poincaré's model of the hyperbolic plane.

Michael Dirda praised Martin's fifty-seven-year retrospective, *The Night Is Large* (1996), as "a mansion of a book in which one can live happily for a month or visit for a quarter-hour, a modern-day equivalent of those catchall classics like Montaigne's essays or Burton's *Anatomy of Melancholy*." *The New York Times* declared *Relativity for the Millions* one of the Books of the Year in 1966 and *Perplexing Puzzles and Tantalizing Teasers* one of the Children's Books of the Year in 1969. W. H. Auden chose *The Ambidextrous Universe* as one of the Best Books of 1965. The English poet described this "discussion of symmetry and asymmetry in the Creation" as "a splendid example of *haute* vulgarisation."

Although his influence was vast and his admirers were countless, Martin never won any major literary prizes, perhaps because his jackdaw intellect resisted classification. He never played the game. He belonged to the Authors Guild until dues got too expensive; he did not feel the organization did anything for him anyway. He did not employ a literary agent. He was not mentioned in *Who's Who*; that was his choice, not theirs. He was rather proud of that glaring omission. He regularly turned down awards if they required him to speak publicly. His peace of mind was shaken when devoted fans introduced the Gathering for Gardner or G4G in January 1993. These conventions held every other year in Atlanta celebrate his work through lectures, performances, displays, and the exchange of puzzles, magic tricks, and other concepts among the converted. Martin attended only the first two sessions. They made him uncomfortable. All the attention frightened him. When the LCSNA offered to meet in Norman to be with him on his 90th birthday, he at first accepted (a very rare thing in itself), then refused to have any part of it. I suspect he was more horrified than flattered when an asteroid was named for him.

Martin Gardner was by nature a modest man. Douglas Hofstadter considered Martin to be "one of the great intellects produced in this country in this century." Stephen Jay Gould called him "a priceless national resource." Martin brushed off any praise that came his way. He was the most self-effacing man I ever met. He hated interviews and only reluctantly granted them. He downright discouraged them. "I don't think my life is too interesting," he confessed to a reporter in 1993. "It's lived mainly inside my brain." Although he demonstrated some magic tricks in film shorts for early television, he was not a public speaker or performer. He had no desire to give lectures or to teach. Fortunately for him, he did not have to. He would never go on a book tour. I suspect signed copies of his books are few. He never inscribed any of his books to me, but he always made sure his publishers sent me the latest one, sometimes in galleys.

He preferred to live unobstructed in relative obscurity and to work at his own pace at home in Westchester County. *The New York Times* summarized Martin early on as "a quiet, soft-spoken man who lives quietly in the suburbs." Charlotte saw to it that he was able to work in peace. She was his most trusted editor. She looked after the household and served as his gatekeeper. She always answered the phone when I called to tell me he was working. Then he would immediately call me back. Being more of a conversationalist, I am a notoriously bad correspondent. Martin did not really like talking on the phone. He much preferred writing. He never bothered to own a Fax or an answering machine.

After twenty-five years, he retired from *Scientific American* in 1981 at age sixty-six. The column was keeping him from writing other books and he thought a younger person should take it over. When he decided to retire to Hendersonville, North Carolina, near the Smoky Mountains, he called me up to tell he was getting rid of a lot of things and invited me to come to Hastings-on-Hudson and take anything I wanted. He gave me some of his Oz correspondence and several precious items he had kept from Jack

Snow's estate. He would have given me anything, had I just asked. Of course he never really retired. He probably was even more prolific once he stopped the monthly column and wrote only what he really wanted to write. "Did he spread himself too thin?" wondered *The Economist*. "It is hard not to think that anyone who writes more books (70-plus) than many people have read, as well as numberless articles and essays, must be at the controls of a sausage machine, yet the sausages were usually good." Some readers actually thought there must be more than one Martin Gardner. How could just one person write so much and so well on such a vast variety of subjects? And all without an advanced college degree?

I learned a lot about writing from Martin. He made everything look so effortless, but this painstaking researcher was an intrepid independent scholar unfettered by academic credentials. He did not owe anyone any favors. "Although my interests are broad," he modestly explained to Kendrick Frazier, "they seldom get beyond elementary levels. I give the impression I know far more than I do because I work hard on research, write glibly, and keep extensive files of clippings on everything that interests me." He regularly sought out those who knew more then he did himself. I often ran into him at the central branch of the New York Public Library at 42nd Street, and he taught me a few research tricks such as diligently scouring reels of newspapers on microfilm for any pertinent tidbit in long-forgotten contemporary accounts and writing to descendants mentioned in obituaries for information. After he moved to North Carolina, Martin hired a mutual friend to aid in his research at the New York Public Library. He was first of all a scientist who demanded evidence. He was great at gathering data. He looked for what other people had to say and explained it to the layman. "Oh, you mean you're in the same racket I am," Asimov once asked him, "you just read books by the professors and rewrite them?"

Martin was a gentleman of the old school. Although fascinated with science, he never mastered the new technology. Always a skeptic, he held his deep doubts about it all. He was not convinced that the computer would ever succeed in duplicating or replacing human thinking. (He likewise challenged the popular notion that chimps and apes could learn sophisticated sign language.) Change is not always progress. He doted on his Xerox machine and was delighted when he finally located a personal microfilm machine. Most of his letters to me were written on an old manual typewriter, with his numerous meticulous corrections in ink. (In the last few years he sent Xeroxes of his correspondence to me rather than the typed originals. I never knew why.) It took his son Jim years to convince him to switch to an electric typewriter. I wonder where they found the ribbons in this digital age! Martin never could write on the word processor. He might surf the Internet, but I doubt he had an email address. I cannot even imagine him bothering with Twitter or Facebook. I have no idea what he thought of e-books, but I can imagine.

I never knew when I might get a quick letter from him either bringing me up-to-date on what he was doing or sharing with me some clipping from the paper or a recent observation of his. Sometimes I received two on the same day! His often brief, succinct notes might open with "A thousand thanks." More frequently they ended with "No need to reply." He never wanted to trouble me. Yet writing him was never a burden and receiving a letter from him was always a joy. There was no end to Martin's generosity to me and others. He had complete faith in anything I might do. He was constantly recommending me to editors right up to the last year of his life. I do not recall ever having to ask Martin for anything: he just offered. He never expected or accepted any thanks. His was pure, unconditional generosity. He was always interested in my opinion, and we frequently exchanged manuscripts to get the other's input. He would send mine back marked up with tantalizing notions that invariably enhanced my arguments. He was meticulous in crediting ideas I shared with him. At times I had to ask him to drop all the constant references to me because it looked like I had made a bigger contribution than I really had. Just an acknowledgement at the end was sufficient. I do not recall him ever saying a cross word to me. He was upfront about what he thought of others and in no uncertain terms, but we never quarreled even if we had some fundamental disagreement.

He also shared with me so many tantalizing ideas for books, often never realized because he could not convince fickle publishers into publishing them. He then would move on. There was always something else to write about. He never lingered. The book Martin told me he was most proud of was his semi-autobiographical novel, *The Fight of Peter Fromm*, about his painful loss of faith, but it took years to find anyone to publish it and even then it was largely ignored by the press and public. It was the novel he had begun and abandoned back in the Chicago of his youth. No one knew exactly what to make of this intellectual adventure. (It seems only one reviewer noted that *fromm* is German for "pious.") That is no wonder: "Disguised as a biography, it chronicles the progressive disillusionment of a young Protestant divinity student at the University of Chicago who, after chucking Christianity, preserves a faith in God." It was not a story of action but one of the mind, a spiritual journey through the philosophical struggle of the changing concepts of God in mid-twentieth-century America. No sex and violence as I recall. Later he changed his mind: his favorite was *The Whys of a Philosophical Scrivener*. It best summarized his thoughts.

When the Oz Club sponsored a contest to publish a new Oz book in 2000 to commemorate the hundredth anniversary of *The Wizard of Oz*, Martin decided it was time he wrote his own Oz story. He updated Baum's 1904-1905 "Queer Visitors from The Marvelous Land of Oz" Sunday comic page by sending Dorothy and her friends to America on the cusp of the Millennium to promote a new movie about Oz. Quickly he realized that he was really writing an adult novel. He thought it would make a great movie. The judges of the Oz book contest did not know what to make of this new kind of Oz story, so Martin sent it to his editor at St. Martin's Press, who published it as *Visitors from Oz* in 1998. Martin sent me an early draft of the manuscript, and I was greatly amused to learn that among the sights the visitors from Oz took in while in New York City was the apartment of a struggling writer named Michael Patrick Hearn! I must have said something to Martin about it for the incident miraculously vanished from the published book.

Martin was proud of his "crazy novel," as he called it, but most reviewers and many Oz fans misunderstood *Visitors from Oz*. It hurt when Michelle Slung in *Book World* judged it "quite simply, a poor thing." *Publishers Weekly* was disappointed by "this rather disenchanted bagatelle, mixing fin de (this) siècle satire with references to several childhood classics." The London *Times* was a bit more ambiguous in calling it "childish stuff and a loving tribute to the original, but strictly for fans (of the books, not the film) only." Some critics found references to the Internet, Möebius strips, and the Klein Bottle baffling and not particularly "Ozzy." What was Dorothy doing on *Oprah*? Perhaps they were expecting a more conventional story in the Baum tradition or something related to the famous movie. On the contrary, it was a rumination on all sorts of things that obsessed Martin over the years, in the form of an Oz book. Taking off from the original satirical "Queer Visitors from The Marvelous Land of Oz" comic page in which Baum commented on contemporary America, it was pure Gardner, not Garland.

Having been brought up in a Protestant household, Martin was never one to wear his heart on his sleeve. He was a professional intellectual: a man of ideas, not feelings. He went to his sister's funeral, sat through the service, and left without saying a word to anyone. The saddest time of his life was when Charlotte died in 2000. He had longed that she might outlive him. At least they were able to greet the Millennium together. I called him immediately when I heard the terrible news. I knew how deeply he loved her and how important she was in his life and work, the most important person he ever knew, but I did not really know what to say beyond offering my condolences. I thought he might just want to talk to someone. He did not sound at all like himself. His voice was uncharacteristically weak, unsure, and it was painful to speak with him at first. I could almost hear the hurt beneath the small talk. Then he opened up. He admitted he was taking medication to get through the depression, through his unbearable loss. He was having trouble writing. For the first time in his long literary life he was suffering from writer's block. He could not concentrate. I feared for him. I told him I would be in North Carolina soon and offered to see him. At first he tried to discourage me and finally *forbade* me from coming. I suspect he did not want me to see him in such a fragile, such a vulnerable condition.

He got through it. His son Jim took control of his life and brought Martin back to Norman, Oklahoma where Jim taught as Professor of Education at the university. On settling into his new life at the assisted living facility in The Gardens at Rivermont, Martin continued to work as diligently and productively as ever. He still had so much to say. In early 2009, *The New York Times* contacted me about an article they were publishing about Oz and I happened to mention Martin. Later they sent me the text that referred to "the *late* Martin Gardner." I immediately wrote back that he was very much alive and that I had just heard from him that week—*twice*! That year he completed the manuscript of his memoirs, "Undiluted Hocus-Pocus: The Autobiography of a Mysterian." One of the last things he ever wrote was an introduction to Baum's *Little Wizard Stories* for Dover, long after he first proposed it. I had edited the book years before for another publishing house and Martin no longer had a copy and wanted to reread my introduction, so I sent him one. Then he sent it back! He did not realize it was a gift.

"People think I'm an atheist, but I'm not," Martin informed me at that very first meeting back in Hastings-on-Hudson. "I have my own beliefs." Skepticism does not strip one of all religious convictions. Martin liked to quote Chesterton: "To an atheist, the universe is the most exquisite masterpiece ever constructed by nobody." Martin never preached, he never sought converts to his religious ideas: he merely stated his position. He said he was belonged to the tradition of Plato and Kant, free of any cant or specific church affiliation. "I decided I couldn't call myself a Christian in any legitimate sense of the word," he explained to Philip Yam in a profile for *Scientific American* in December 1995, "but I have retained a belief in a personal God. I admire the teachings of Jesus, but to me it's a bit dishonest if you don't think Jesus was divine in some special way." Nevertheless, "I think my beliefs are much closer to Christ's than those held by some Protestant ministers I have met." He was a "philosophical theist." "I believe in a personal god, and I believe in an afterlife, and I believe in prayer," he told Don Albers in *The College Mathematics Journal* in 2005, "but I don't believe in any established religion." He was perfectly happy to let other people believe whatever they wished. He was generally a generous thinker. "I am quite content to confess with Unamuno," he revealed in *Whys of a Philosophical Scrivener*, "that I have no basis whatever for my belief in God other than a passionate longing that God exists and that I and others will not cease to exist. Because I believe with my heart that God upholds all things, it follows that I believe that my leap of faith, in a way beyond my comprehension, is God outside of me asking and wanting me to believe, and God within me responding." For Martin, "It is a way of escaping from a deep-seated despair."

The last letters we ever exchanged happened to concern, among other topics, our personal religious beliefs. In referring to the recent death of a mutual Oz friend, Martin wrote Fred Meyer back in 1960, "it suddenly occurs to me that part of the fascination of Oz lies in the fact that it is surrounded by the desert, a kind of death symbol that fits in beautifully with the immortality of Oz inhabitants. Once over the desert into Oz, and no more deserts to cross." The last thing Martin ever wrote me was

"I'm delighted to hear you're not an atheist. I recently had a pleasant visit from Dawkins, of all people!" British biologist Richard Dawkins wrote the international bestseller *The God Delusion* (2006), in which he argued that any belief in a personal God was absurd. He considered Martin Gardner to be "one of the great heroes of the American sceptical movement." Martin slyly ended his letter to me, "We agreed to disagree about God."

I learned of Martin's death, in the Norman Regional Hospital on May 22, 2010 after a brief illness, so coldly, so bluntly, from an unsolicited email message. Of course it was not entirely a surprise: he was 95, after all. Yet I believed he would surely reach 100. At least I *hoped* he would. He still seemed so vigorous, so focused in his letters. He had so much more to say. It was no surprise that he *forbade* any memorial service. In his humility, he would not have approved of any festschrift in his honor, and would not have read it. Although I never told him, I had every intention of dedicating my next book to him. I wanted it to be a surprise. I have since changed the manuscript from "For Martin Gardner" to "In fond and grateful memory of the Master, Martin Gardner."

Martin Gardner, Major Shaping Force in My Life

DOUGLAS HOFSTADTER

I've been trying to reconstruct how I first encountered Martin Gardner. It may have happened in 1959, when at age fourteen I happened to visit the home of a boy a couple of years older than myself, who I thought was extremely smart (and indeed he was—he later became a well-known mathematician on the Princeton faculty). While scanning his bookshelves, I noticed a Dover paperback with the curious title "Fads and Fallacies in the Name of Science". I pulled it out and my curiosity was further aroused by the front cover, which mentioned such things as flying saucers, human gullibility, strange cults, pseudoscience, and so on. I had of course heard of things like telepathy, ESP, and such, but didn't know what to make of them. Though they seemed a bit farfetched, they also appealed to my romantic nature. The year before, I had even half-convinced myself that I could discover my romantic fate by spinning a top and seeing where it fell

Douglas R. Hofstadter directs the Center for Research on Concepts and Cognition at Indiana University in Bloomington. In 1981, when Gardner retired from writing his column for Scientific American, *it was Hofstadter who succeeded him, with a column entitled "Metamagical Themas", an anagram of "Mathematical Games". Hofstadter's book* Gödel, Escher, Bach: an Eternal Golden Braid, a metaphorical fugue on minds and machines in the spirit of Lewis Carroll *(1979) won both a Pulitzer Prize (General Nonfiction) and an American Book Award (Science). His other books include* Le Ton beau de Marot: In Praise of the Music of Language *(1997), and* I Am a Strange Loop *(2007).*

on a marked board; I also enjoyed the thought that maybe, just maybe, the first initial of the girl I would someday marry would be revealed by reciting the alphabet as I twisted an apple stem and stopping at that letter when the stem broke off. Why not? At that tender and rather gullible age, I had never devoted much thought to the demarcation line between sense and nonsense, science and silliness.

But in this book, somebody—clearly somebody very intelligent—was tearing one oddball belief system after another to shreds in a lucid, acerbic, yet at the same time humorous way. This "Martin Gardner" person was wielding common sense as a surgeon wields a knife—and occasionally twisting the knife with glee. It was probably the first time I had realized that systematic and critical thinking could extend beyond such precise domains as math and physics, and could demolish ideas in far hazier fields with great power. It was also the first time I had realized how very many crazy belief systems there are out there in the world, and how important it is to recognize this fact and to combat them.

Something about that book impressed me deeply. When I bought my own copy and read the whole thing, my ideas about the so-called "paranormal" were forever changed. In his preface to the second edition, Martin observed that, of the many readers whose ire had been aroused by the first edition and who wrote to tell him so, most focused their venom on the one chapter that attacked their pet belief system, while curiously enough thinking the rest of the book excellent. This wry little remark said volumes about human nature, the nature of truth, and the nature of objectivity. I've never forgotten it.

Fads and Fallacies may have been my introduction to Martin Gardner, but another avenue strikes me as equally plausible. My family had subscribed to *Scientific American* forever, it seemed, and I may well have encountered the monthly "Mathematical Games" column in browsing through an issue. In any case, I vividly remember the day when my mother brought home for me *The Scientific American Book of Mathematical Puzzles & Diversions*—Martin's first collection of his columns. Reading through that book was a mind-opening experience for me, a teenager deeply in love with math.

Martin had a magical touch in writing about math. His column's title was very humble and in a sense misleading. The word "games", with its lightweight flavor, did not even hint at the depth of the issues that the column dealt with. Theoretically, it was about "recreational math", which sounds frivolous, but in fact the column was about beauty and profundity in math—and in many other fields as well. In each column Martin managed to put his finger on some little known but profound issue, and to present it in such a clear (and often droll) fashion that its importance just grabbed me and infected me like a virus. After reading a Gardner column, I would often walk around for days with the ideas swimming through my head, like an incredibly catchy melody.

Just to mention a few, there were the Soma cube, the inductive game of Eleusis, 4-D tic-tac-toe, nontransitive dice, the unexpected-hanging paradox, Newcomb's paradox, nonconservation of parity, machines that learn, thinking machines, squaring the square, fallacies in mathematical reasoning, cryptography, knot theory, Nim and its variants, properties of e and π, random walks, tesseracts, hyperspheres, Fibonacci numbers, Lucas numbers, Catalan numbers, Pascal's triangle, numerological nonsense, Carl Hempel's "all crows are black" paradox, Nelson Goodman's bleen-and-grue puzzle, palindromes, Möbius strips, Klein bottles, spirals, helices, cycloids, conic sections, the paradoxical art of M. C. Escher, communication with extraterrestrial intelligence, infinity, isomorphisms, graph theory, pseudo-random numbers, curves of constant breadth, superellipses, 4-D space, Zeno's paradoxes and supertasks, the $3n + 1$ problem, the game of Life, Fermat's last theorem, mathematical music, magic squares, the four-color problem, time reversal and time travel, Gödel's theorem, topological games, knotted holes—and on and on and on it goes. In truth, I have just scratched the surface!

Lurking behind it all, it seems to me, was Martin's passion for paradox. I would say that more than anything, this passion gave and continues to give to Martin his virtually unerring sense for what is important.

Paradox comes in many forms. Special relativity violates our intuitions about time and space at such a basic level that, even though it is far from being self-contradictory, it seems like absolute nonsense when one first encounters it. Four-dimensional geometry, likewise, posits an unvisualizable space—within which one then gaily plays! The chaos of the prime numbers is almost a paradox—at least it is a deep mystery, and it can provoke and inspire thoughts for as long as one wishes to think about it. Cutting a Möbius strip down the middle and not getting two pieces is as crazy an experience as one could ever wish to have!

How can it possibly be that "All crows are black" and "All non-black things are non-crows"—two sentences that logically are exactly equivalent—have completely different meanings when it comes to verifying them by citing examples? How can a sentence be readable backwards as well as forwards? That's nonsense! What is the unfathomable secret to the power of music, and do the never-ending patterns upon patterns of mathematics have anything to do with it? Why are the fundamental laws of nature symmetrical—or are they? What would happen if time started to go backwards, or left and right were suddenly switched? Why does a mirror seem to reverse left and right but not up and down? If infinity has no end at all, how can there be different sizes of infinity? How could a tabletop roll completely smoothly on a set of logs all of which have different cross-sections, none of which is circular? It all makes no sense. It simply makes no sense. At least at first sight.

Again, I could go on and on, but the point is that Martin's style is intimately bound up with paradox and mystery—the clear exposition of paradox, but just as much, the clear resolution of mystery whenever possible.

Martin's columns and writings radiate a profound exuberance in the constant novelty of human thought. What comes through, even if it's never explicitly expressed, is a kind of informal version of Gödel's theorem for human thinking—a sense that creative minds will always one-up the pedestrian expectations generated by unimaginative, logic-bound thinking. There is an exultation in the breaking-out of expected patterns, the violation of seemingly ironclad laws, the making of wildly unexpected connections, the revelation that two seemingly identical properties are really quite different, and the counterexamples that make it all blindingly clear (at least for a moment—then you forget how it worked!). . . . If nothing else, reading Martin Gardner should convince you that the human mind's pathways of finding truths are as diverse and unpredictable as the pathways of evolution itself.

Not surprisingly, mixed in with this celebration of the unpredictable was a love for humor and oddities, which are certainly close to the antipodes of pure logic (as is, in a sense, Gödel's theorem). "Mathematical Games" was always bringing to its readers such diverting items as poems written without the letter "e", limericks with the wrong number of lines, astonishing anagrams and numerical coincidences, droll poems on science and math, endless new varieties of wordplay, paradoxical pictures and sculptures, surprises of figure–ground play, and so on. And in April issues, Martin would occasionally try to pull hoaxes on his own loyal readers, to test their gullibility.

Even those columns simply labeled "Nine Problems" always had something classical and fine about them. Never was a problem a mere exercise without a deeper point. Every one of them contained a lesson in how to think. Once again, common sense was being put to the test in ever novel ways.

I remember that when I was a teenager, every time I discovered an issue of *Scientific American* in our family's mailbox, I would instantly flip it open to the "Mathematical Games" column, filled with excitement and curiosity to see what amazing new set of ideas would be discussed. Some of my friends felt exactly the same way, and many years later I came to understand that there were thousands of such people spread all around the world—mathematicians, physicists, philosophers, computer scientists, and on and on—who thought of Martin Gardner's column not as merely a *feature* of that great magazine *Scientific American,* but as its very heart and soul.

In the early 1970s, when I was a graduate student in elementary-particle physics, I went through a period of deep crisis. I found myself utterly unable to relate to any of the ideas I was exposed to through readings and seminars. The more I thought about my future, the gloomier it looked. There came a point where I could see no future for myself other than teaching math or physics in a high school or junior college. Though I loved teaching, this would mean foregoing research, foregoing the pursuit of novel ideas, foregoing creativity. And this didn't seem at all right for me.

After all, despite my deep discouragement, some part of me was convinced that I had a creative mind, even a mind with something of the Martin Gardner spirit. I had always resonated deeply with the paradoxes that he offered, and more than once I had come up on my own with something that he later published in his column, submitted by somebody else. In my files, I had amassed a sizable collection of quirks and oddities and mathematical surprises, and in my greatest flights of fancy, I would imagine being offered the chance of writing the "Mathematical Games" column myself, if Gardner should ever retire. But it felt like an utter pipe dream. How would the higher-ups at *Scientific American* ever come to think of *me,* a lowly physics graduate student, as a worthy successor to Martin Gardner?

And yet, just a few years later, that exact dream did come to pass. I never did reconcile with myself with particle physics, but instead moved into solid-state theory, and got my Ph.D. in it. However, as soon as I had done so, I slid out of physics and into artificial intelligence, sensing that my deepest interests were actually in unraveling how the mind works. Those interests, interacting with many of my other interests, gave rise to my first book, *Gödel, Escher, Bach: an Eternal Golden Braid*, to which Martin devoted a whole column in *Scientific American* and on which he lavished praise. I was and am enormously indebted to Martin for that act, which probably more than anything else made my book known and launched it on a pathway of success. It was quite a thrill to be touted in the very column that I had so greatly admired and had followed so faithfully for so many years.

Only a year or so later, Martin decided to stop writing his column in order to have more free time. Could someone be found to carry on the "Mathematical Games" spirit? I believe it was Martin himself who suggested to *Scientific American*'s editor and publisher, Dennis Flanagan and Gerard Piel, that I might be a plausible person to consider. When Flanagan and Piel approached me with this thought, I was both overwhelmed and frightened. I had in the meantime become a professor of computer science and was seriously engaged in artificial-intelligence research. How could I continue to do my research and also do justice to the column that Martin Gardner had created, which by then had turned into an international institution?

After churning it over in my mind for several weeks, I finally decided to risk it, because it simply was too good an opportunity to turn down. I knew that I would forever regret it if I failed to grasp the very chance that I had once yearned for and that had seemed beyond my wildest dreams. Moreover, in a letter to me, Dennis Flanagan said explicitly that I should feel free in following Martin's lead in writing about *anything under the sun that interested me*—that it need not concern either mathematics or games, that it could be scientific, literary, artistic, what-have-I. Talk about carte blanche! It was almost too good to be true.

Nonetheless, I felt trepidation about being placed precisely in Martin Gardner's shoes. Even though I felt we were in some ways kindred spirits, I also recognized that our interests and skills were not by any means identical. It would give readers a false set of expectations if I simply adopted Martin's mantle, and it would put a heavy set of burdens on me. So I tried to preserve the spirit but not the letter of Martin's column by making the title of my own column an anagram of his title. I took the seventeen letters in "Mathematical Games" and scrambled them to make "Metamagical Themas". By this gesture, I tried implicitly to let readers know not to expect me to be Martin Gardner, but that I would try to carry on some of his marvelous spirit.

In order to prepare for writing my column, I felt it would be a good idea to meet Martin Gardner, for he was the person whose mantle I was taking over, the person who had set the magical tone of the column for some twenty-five years. And Martin in turn was very glad to welcome me and give me some friendly "coaching". And thus, sometime in the fall of 1980, I visited him in his home in Hastings-on-Hudson, New York, and met his extremely intelligent, extremely generous, but no-nonsense wife Charlotte. I will never forget the bell up on the third floor of their house, where Martin had his office and where he and I sat and talked for hours. Down on the first floor, Charlotte tugged a string and the bell rang to summon Martin and me down to lunch—and no dawdling was allowed.

After lunch, Martin and I climbed the stairs to regain his office, and at some point, while we were talking about the challenges of writing a monthly column with so many readers of such brilliance, and also about our many overlapping interests, the phone rang. When Martin answered it, I heard him talking with someone first about mathematical logic and then about Zen Buddhism and Taoism. To my amazement, this turned out to be another long-time hero of mine, Raymond Smullyan, whose book *Theory of Formal Systems* had had a big impact on me as a teenager, and who had just finished writing a new book called *The Tao is Silent*. Smullyan's multiplicity of interests really struck me, and all the more so when I found out from Martin that Smullyan was also a top-notch magician and an excellent pianist.

I realized then that Martin was a kind of hub for communication among an amazing set of sparkling intellects in all sorts of disciplines.

Some of the brilliant people whom Martin linked together were the statistician/magician Persi Diaconis, the mathematician/juggler Ron Graham, the Uri Geller–debunker and psychologist Ray Hyman, the mathematician/logician/musician/magician Raymond Smullyan, the magician and pseudoscience-fighter James Randi—and there were many, many others. It was humbling to realize that this very unassuming, gentle man sitting across the table from me was so admired by, and so *central* to, so many people who were gifted with both extreme intelligence and extreme creativity.

Well, soon my column began, and I had a grand time for some three years, writing on all sorts of topics, some that I knew Martin himself would surely have written about had he not retired (such as Rubik's cube), others that I knew were quite unlike his own intellectual directions (such as sexism in language). But the burden on me grew greater and greater, and eventually it became clear that I would be unable to continue to produce columns at a monthly pace. My column thus wound to a close in 1983, which freed me up to return to my research. Luckily, just as Martin had done, I was able to take my pieces and convert them into a book. I had had a three-year affair with a dream, and then it broke up—but it certainly was great while it lasted. I have since run into quite a few people who, just like me at one time, would have given their eye teeth to be in Martin Gardner's shoes. I had been really lucky!

Since retiring from the "Mathematical Games" column, Martin has continued with full vigor to pursue all of his far-flung interests. In fact, I cannot understand how he finds the time to read and write as many diverse books and articles as he does. In the past twelve years, his writings have deeply enriched the mathematics community, the anti-pseudoscience community, the magic community, the wordplay community, the philosophy community, the science-fiction community, and on and on.

Even though Martin is very self-deprecating and would disagree with this, I feel that an intellect like his is a treasure. Many of today's most influential mathematicians and physicists, magicians and philosophers, writers and computer scientists owe their direction to Martin Gardner. They may not even be aware of how big a role he played in their development. After all, influence is often transitive—if A influences B and then B influences C, A may thereby have a large influence on C, yet C may never have even heard of A!

There should, it seems to me, be a prestigious national or international prize for writing about scientific ideas. As everybody knows, human civilization relies on science and technology more than at any time in the past, and that reliance can only increase. Yet the worldwide ignorance of and disdain for science, mathematics, and precise thinking in general is appalling. Because of this tragic situation, people like Martin are precious purveyors of precious knowledge. I hesitate to use the word "popularizer" because, to me, that word has always carried a connotation of being subordinate, derivative, and secondary—in short, of lesser importance than the originators of scientific ideas. And in many instances, this is undoubtedly the case. But in Martin's case, it is emphatically not. His approach and his ways of combining ideas are truly unique and truly creative, and, if I dare say so, what Martin Gardner has done is of far greater originality than work that has won many people Nobel Prizes. Simultaneously achieving both depth and breadth is almost unheard of in today's scientific world, but Martin Gardner is an exception, and it is a delight and a privilege to celebrate here his many achievements. Just as Martin's writings have inspired me for decades, so they will undoubtedly continue to inspire other people for many decades to come.

POST SCRIPTUM, 23 MAY 2010

This is really a sad day. Not so much sad that Martin died, since we all knew it had to come pretty soon, but sad because his spirit was so important to so many of us, and because he had such a profound influence on so many of us. He is totally unreproducible—he was *sui generis*—and what's so strange is that so few people today are really aware of what a giant he was in so many different arenas of the mind. To name some of them: the propagation of truly deep and beautiful mathematical ideas (not just "mathematical games", far from it!), the intense battling against pseudoscience and related ideas, the invention of superb magic tricks, the love for poetry's traditional musical beauty and the lamenting of its fading, the fascination with profound philosophical ideas (Newcomb's paradox, free will, etc.), the elusive border between nonsense and sense, and the idea of intellectual hoaxes carried out in order to make serious points. For example, one time, at my instigation, Martin wrote a long and scathing review of his own book *The Whys of a Philosophical Scrivener* in *The New York Review of Books,* the idea being that, on its surface, the review would angrily slash the book's ideas left and right, but if read on a deeper level it would reveal all the weaknesses of opposing views, and thus in the end, the review's harshness would serve the book well—and Martin penned this clever self-attacking review under a pseudonym ("George Groth") that he wittily unmasked in the review's very last sentence.* What a delight!

For all his influence on so many top-notch thinkers in so many disciplines for so many years, Martin Gardner was impeccably modest and unassuming. So it is sad to think that such a person is gone, to remember that so many of us owe him so much, and to realize that so few people today—even extremely intelligent, well-informed people—are aware of who Martin Gardner was, or have even ever heard of him. Very strange. But I suppose that when you are a total non-self-trumpeter like Martin, that's what you want and that's what you get. And so perhaps it's all for the best that he remains mostly hidden behind the scenes, known only to a special set of people.

It is a very sad piece of news, but Martin lives on in us—in so many of us—and he will do so for a very long time. This is certainly a time of great sorrow and a time to reflect back on how important he was in forming us. RIP, dear MG.

* Reprinted as "Gardner's *Whys*" in *The Night is Large: Collected Essays, 1938–1995* (St. Martins, 1996) – Ed.

✾ This article was first written in 1992 and presented by the author at the first Gathering for Gardner. It was posted on *Scientific American*'s website shortly after Mr. Gardner's passing in 2010, but has not before appeared in print. As Gardner was very much alive at the time of its writing, we have retained the present tense. For this volume, Dr. Hofstadter has added a postscript based on emails he wrote the day he learned of Gardner's death. At the author's request, we have retained his punctuation style.

The Stanford Gardner Archives

STANLEY S. ISAACS

Martin Gardner kept many files on recreational mathematics. During the winter of 1994, Donald Knuth visited him and "lived with the files" for three weeks. He went through all the files from the "Mathematical Games" column, making notes for his book *The Art of Computer Programming*. He said it was such a treasure trove of information that it was "like discovering King Tut's tomb."

When Gardner was deciding what to do with his library and files in preparation for a move to smaller quarters in a different state, Knuth thought of the possibility of archiving them at the Stanford University Libraries, where Knuth's own files are preserved. He brought up the idea of obtaining the files with Persi Diaconis, a close friend of Gardner's, and Margaret Kimball, the University Archivist of Stanford University. Kimball, in turn, talked to Henry Lowood, Curator of the History of Science. Everyone agreed that it would be a good fit to acquire the files for Stanford, and Diaconis (and Knuth) convinced Gardner that the Stanford University

Stan Isaacs has been collecting books by and about Lewis Carroll for over thirty years, since he inherited his grandfather's Carroll collection; Stan's emphasis is on mathematical works, translations, and the many illustrators. In 2002, he retired from fifty years of being a computer programmer/analyst, the last twenty at Hewlett Packard. He currently teaches computer classes at the College of San Mateo while he continues to avidly pursue his avocations of mathematics, researching and teaching vintage ballroom dancing, and both collecting and playing with mechanical puzzles.

Libraries would be an ideal place for his papers. And thus, in 2003, about 20 large, deep, 4-drawer metal file cabinets, plus 31 boxes of books, were delivered to Stanford University.

Martin Gardner had many files, reflecting his many interests. In addition to the files on mathematical recreations, he had files on philosophy, on magic, on Lewis Carroll and *Alice in Wonderland*, on Frank Baum and *The Wizard of Oz*. He had an extensive history of science collection. He had over 500 books of poetry. He had a large collection on psychic phenomena, which he donated to Committee for the Scientific Investigation of Claims of the Paranormal (CSICOP) in Buffalo, New York. The files went various places, but the bulk of the mathematics-based files—recreational and otherwise—came to Stanford.

The contents of the files, over 2,000 folders, were moved into boxes to be indexed and protected. I worked as a volunteer in the Department of Special Collections for three years to prepare the contents of these boxes for archiving. My job was to put the folders into more accessible form, in smaller archival boxes with about 15 to 20 archival folders per box, put delicate papers into plastic, photocopy newspaper pages, etc. I also entered the basic information about each folder in a FileMaker Pro database, so that they could be made available to all researchers. The indexing is fairly cursory, and not at all complete. I hope, now that this first step has been achieved, someone will obtain a grant and go through all the files in detail and make a more complete index. But Gardner himself left it to us well organized and with excellent folder titles, which we have retained. The best overall summary of the files are in the series of books he wrote collecting the *Scientific American* columns and addenda with new material from the files. Every time a new edition came out, he added appropriate new material.

The folders contain an amazing variety of material. There are lots of Gardner's notes: ideas for columns, working out problems, summaries of articles, etc. There are many letters from correspondents, almost always with a copy of his answer or notes about it on the letter. He often served as a conduit between people of like interests—whenever someone wrote to him on a particular subject or problem, he was able to put that person in touch with others who were working on similar problems. There are also preprints, reprints, typescripts, and first drafts of articles, mostly by others, rarely by Gardner himself. Most of the *Scientific American* columns themselves are there, but not all. There are very few of his non–*Scientific American* articles, nor his early drafts of them. There are occasional flat models, photos, drawings, games, etc. from his correspondents; many magic trick descriptions, both from Gardner himself, and from others; and a lot of original artwork.

I found the papers fascinating. One of the reasons it took so long for me to prepare the papers was that I kept finding interesting things to read. Much as I tried *not* to read too much, it was difficult to avoid getting hooked into reading interesting items, notes, letters, pictures, problems, etc. As a puzzle collector, I kept finding puzzles or analyses of puzzles I hadn't seen before. As a Lewis Carroll collector, I kept seeing references to mathematical puzzles or books that Carroll wrote, as well as to the *Scientific American* columns devoted to Carroll. One of the things that had started me long ago in puzzle collecting was a Gardner article on mechanical puzzles. I came across pictures of the puzzles in Lester Grimes's collection, from a September 1959 article. I also came across folders on sliding block puzzles, the Soma Cube, peg-jumping puzzles, M. C. Escher, and more. Another group of folders of particular interest were the ones about Douglas Hofstadter, which include several summaries of his book *Gödel, Escher, Bach*, and a long analysis by Scott Kim.

Gardner was particularly influential, not only because of his great expository writing style, but also because of his organizational skills and his diligence in answering every query quickly. Central to this discipline were his files. Anything that crossed his desk was read, frequently underlined, and then indexed and filed.

To access the files today, one needs to contact the Stanford University Libraries. From the Web page at `library.stanford.edu` you can find instructions for requesting materials: online, in person, or by email.

To search Stanford Library's online catalog, go to `searchworks.stanford.edu`. Select `Special Collections & University Archives` from the `Location` list at left. On the resulting page, put `Martin Gardner` in the search box and click `Search`. From the results box, click `Papers, 1957-1986`. This will bring you to the heading page.

If you are interested in going to a searchable database of the collection, in the upper left, under `Related E-Resources`, click the link to `www.oac.cdlib.org`, where you will be taken to the `Finding Aid`, which is the index of boxes and folders. The titles and dates of all the folders are there, as is a link to a list of correspondents and which folders the letters are in.

If you wish to physically look through the boxes and will be at the Stanford's Special Collections and University Archives reading room, there is a list of boxes, e.g., `SC 647 SERIES 1 BOX 7`, with a clickable `[request]` link next to each, so that requests can be submitted online in advance of your visit (you will need to make requests at least 2 to 3 business days in advance). You are limited to 5 boxes at a time.

The collection is organized in two series. Series 1, the Game Files, consist of 56 boxes containing the folders related to the "Mathematical Games" columns that appeared in *Scientific American* between the years of 1956 and 1986. The titles of the folders retain the headings that Gardner used. These do not necessarily correspond to the chapter titles in the books of collected columns since the columns did not have titles when originally published in *Scientific American*. (The bibliography at the end of this volume will be of help.) But they are in order by the date of the column, and the first folder for each column has the date as its title. Most columns are there, but occasionally there is just a pointer to a later folder. The dates are very rough, sometimes including dates of articles, but usually not. There are many letters, and I've tried to list the correspondent for most of them, especially if they contained an interesting analysis or were from well-known people. But many were just signed with first names, or unreadable last names, and these were not usually included in the index.

Series 2 are 75 boxes of the Subject Files, which are arranged by topic. Some of the topics are very general, like plane geometry or combinatorics; others are more specific, like serial isogons, or MacMahon's color polygons. They tend to contain more articles and Gardner's notes and fewer reader letters than the column folders, though there are many of each in both sections.

The Gardner files at the Stanford University Libraries are limited to his files on recreational mathematics. He had a similarly voluminous archive of Carroll material; its whereabouts are unknown at this time and we are actively trying to determine where it is.

Preparing the Martin Gardner papers for archiving at the Stanford Libraries was a labor of love. But I hope that the papers will prove to be as interesting and exciting to everyone else as they have been to me.

Martin Gardner: Through the Looking Glass

SCOTT KIM

Yes, Gardner wrote *The Annotated Alice*. But the connection between Martin Gardner and Lewis Carroll is deeper and curioser than you might suspect. Let me explain.

As a child, I read puzzle books to fuel my passion for mathematics. In high school I wrote a postcard to my favorite math book author, Martin Gardner, telling him about my attempt to solve Fermat's Last Theorem. He wrote back promptly, stating that while he was not qualified to evaluate my mathematics, he applauded my effort, and recommended a book to look at.

I was too young to have known at the time, but this should have made me suspicious. I later learned that everyone who ever wrote to Martin received a prompt, hand-typed reply. How could a celebrated author have the time to respond to every letter? Furthermore only a deep mathematician could have written his books, and mathematicians receive far too many letters claiming to have proved Fermat's Last Theorem to respond with anything

Scott Kim is an American puzzle and computer game designer, artist, ambigramist, and author. He has been a columnist for Discover magazine for many years, and has created hundreds of puzzles for magazines such as Scientific American, Wired, *and* Games, *as well as for the Web, computer games, math fairs, games and puzzles conferences, educational use, and toys. Isaac Asimov called Kim "the Escher of the alphabet."*

more than a form letter. I began to suspect that Martin was not, perhaps, who he claimed to be.

During my college years, I visited Martin at his quaint home in Hastings-on-Hudson. I showed him my latest work on ambigrams,[1] and he showed me the column he was writing, about the Italian puzzle sculptor Miguel Berrocal. Responding to my ambigrams, he went to his files and pulled out this gem from his folder on numbers:

3.14 in a mirror is **PI.E**

I spent a pleasant afternoon rummaging through his voluminous filing cabinets. There I found manila folders on every column he had ever written, including fascinating material on my then-current obsession, the fourth dimension. Every letter he had ever received was there, neatly filed with his response, and entries were often photocopied so they could be cross-catalogued.

Such a superhuman level of organization again raised my suspicions. My puzzle friend Stan Isaacs, who catalogued Gardner's files for Stanford University, confirmed that the files were left in perfect order. Stan has not had to re-file any misplaced papers. And like me, Stan simply gets engrossed in the material.

Several years later Gardner published an article about my ambigram book, *Inversions*,[2] in his *Scientific American* column "Mathematical Games." To my ambigram designs, he added his own hilarious symmetrical word, in which the variety of strokes is kept to a MINIMUM:

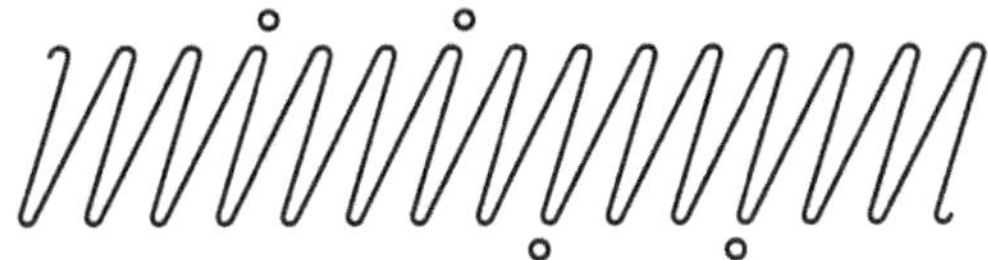

Appearing in Gardner's columns helped launch my career as a puzzle designer and puzzle columnist. For this and for many pleasant visits I am grateful. But who is Martin Gardner, really?

After years of scientific investigation, I have concluded that "Martin Gardner" is a pseudonym, much like "Bourbaki," the name of the prolific French mathematician who turned out to be a team of them. Martin's prolific correspondence, voluminous publication, and fastidious organization simply could not have been the work of one man.

The name itself has many remarkable properties, which can hardly be a coincidence. For instance, it can be written in rotational symmetry,

martin Gardner

in mirror symmetry,

Martin
Gardner

as a pentagonal star,

or as an interlocking pattern on the surface of a cube.

Furthermore MARTIN GARDNER can be written so upside down it reads as DOCTOR MATRIX, Gardner's most famous fictional character:

Clearly Gardner's name was deliberately fabricated. Want further proof? Rearrange the letters of MARTIN GARDNER and you get MANTRA GRINDER, surely a reference to his work as a skeptical critic of pseudoscience. His *Scientific American* column MATHEMATICAL GAMES rearranges into METAMAGICAL THEMAS, the title of the column by his successor Douglas Hofstadter, another suspicious character.

But if Martin Gardner was not who he claimed to be, who was he really? One only needs to rearrange the letters of his name to reveal the phrase MAD ERRANT GRIN, a reference to Alice's Cheshire Cat. And if that is not evidence enough, turn the name below upside down.

[1] An ambigram is a calligraphic or typographic design that may be read in more than one way, as one shifts viewpoint, direction, or orientation. The earliest known ambigram dates to 1893 by artist Peter Newell, whose 1901 illustrations to Carroll's books were used by Martin Gardner in his *More Annotated Alice*. The last page in Newell's book *Topsys & Turvys* contains the phrase THE END, which, when inverted, reads PUZZLE (see p. 234 – Ed.).

[2] Byte Books/McGraw-Hill, 1981; repr. Key Curriculum Press, 1996.

Martin Gardner & Annotation

JIM KINCAID

Martin Gardner, doubtless the best known and most widely respected among annotators of our time, was also the kindest, by a wide margin. I'll get to his work as an annotator shortly, but first let me deal with the kindness. In some walks of life—one thinks of saints, martyrs, or (not to be immodest) certain college professors—kindness might be seen as an occupational given; but no one who knows them would place annotators in this class. A rancorous and waspish lot they are, almost without exception. But we are dealing with an exception here—perhaps the single exception—so it's well to add that he was also a famous Carroll scholar and annotator. As we all know, Carrollians are marked almost universally by their disagreeable nature and habits; so Martin Gardner's position is doubly remarkable.

"How do I know this?" you ask. I'll tell the story, annotating as I go, as I am myself one of those malevolent and best-avoided-if-possible annotators.

A student of Victorian literature and society and literary theory, Professor Jim Kincaid holds the Aerol Arnold Chair in English at USC. His published works include the preface and notes to the Pennyroyal/University of California Press editions of the Alice duad (1982), Child-Loving: The Erotic Child and Victorian Culture *(Routledge; 1992), and* Annoying the Victorians *(Routledge, 1994). Professor Kincaid hosted the "Lewis Carroll and the Idea of Childhood" conference (conflated with our Society meeting) at the University of Southern California and the Huntington Museum and Library in 2006.*

At the sesquicentennial celebration (1982) of Dodgson's birth, sharing the podium at The Pierpont Morgan Library with Morton Cohen and other notables, I gave a talk to the Lewis Carroll Society of North America, "Confessions of a Carroll Critic," later published to wide acclaim—ha! It would be unbecoming of me to characterize that talk as, say, "brilliant," but I don't know that it deserved the clear contempt with which it was met. There were two exceptions to this deep-freeze, however: my wife* and Martin Gardner. My wife is ready to offer testimony on the talk's excellence at any time, but only Martin Gardner took the time to pen an appreciative note and then—wow—to cite the talk and some of my annotations† in his dazzling *More Annotated Alice* (1990).

But what sort of annotator was he? First, one must say that he was both experienced and acclaimed, not just for the work on Carroll, which is my focus, but of G. K. Chesterton, "Casey at the Bat," "The Night Before Christmas," and "The Rime of the Ancient Mariner." As a distinguished philosopher, popular scientist, and recreational mathematician (a delightful field he championed), Gardner filled his annotations with fascinating material from these fields, as only he could.

But we are after something deeper—or wider—or at least different. Annotation is very far from a subservient practice or an innocent one: It does not take a back seat to the text it pretends to be supplementing but offers instead a rival text, competing for our attention while sneakily pretending to do nothing more than assist our reading of something else, *Alice's Adventures in Wonderland,* say. But I've had my poststructuralist say on that point in the "Confessions" essay I have no reason to suppose anyone, apart from the two exceptions noted, has read; so I'll move on.

Annotators differ widely in their practice but still fit neatly into three distinct categories: the scholar, the sprawler, and the sneaker. As illustrative of "the scholar" category, one might cite almost any classroom edition of a Shakespeare play, or, indeed, about 99% of annotating practices. These fellows and gals can be known by their works: those seemingly servile, self-effacing notes which pretend to be dictated by the text itself, in cooperation with the assumed ignorance of the reader that calls into being our helpful

* Long-suffering, she says, and ill-rewarded.

† From my notes to the Pennyroyal/University of California Press *Alices*, designed by Barry Moser.

annotator. Were we able to read these words, phrases, structures of meaning on our own, annotators would be out of business; so they presume to come to our aid, armed with their superior historical, lexical, biographical, cultural knowledge. It's all a sham, of course, as these annotators are show-offs and bullies in disguise, letting us think that what they happen to know (or have just looked up) is what's essential. But I don't want to be too severe on scholar-annotators, as those mendacious traits they demonstrate are less character defects than qualities native to the craft.

The sneak category is represented by—well, me. Under cover of the darkness accorded to margins, I smuggled into the Pennyroyal and California *Alices* a running interpretation of the works, in no way essential to anyone's understanding and probably, in most cases, a positive roadblock. I felt that these works could be read as partly hostile to their child-subject, covert attacks on Alice for being so very adult, so very obtuse, and thus unwilling or unable to understand the attractions of the anarchic adventure held open to her. Underneath all, Alice is seen as the child who would grow up, the antithesis of Peter Pan, refusing the proffered love of the White Knight and of Carroll. And so forth. It would be hard to argue that anyone needs to hear such things, much less that they can legitimately be offered as a gloss on what Carroll wrote. Very true. Sneaky is what it is.

The best form of annotation, the one represented with most distinction by Martin Gardner, is the sprawler. He would not disagree. In his gloss on Alice's punning reference to having no room to grow (while her mushrooming body is filling up the house), Gardner says, "I thank Mr. Kincaid for supporting my ramblings." That's kind, as I mentioned; but who would not support such ramblings as we are given by Martin Gardner on virtually everything in the *Alices* on which it is possible to ramble?

Gardner is, as expected, brilliant on all puzzles and games, on all puns and problems. My favorites, though, are those of his annotations dealing with science—speculations on actually falling through the earth to the other side, for instance—and on matters of taste. Gardner has a ill-concealed aversion to Freudian readings, usually expressed in sneers: "as Freudian critics never tire of telling us." He also does not hesitate to tell us what he thinks of things such as the 1999 NBC televised version in which Whoopi Goldberg starred: "undistinguished, boring." He also loved animals—that's a guess but supported by his glorious comments on ferrets, which goes on for—oh—

hours, winding toward news of today's attitudes toward ferrets: not legal in NYC, as health-codes prohibit them, but promoted (as they damn well should be!) by some enlightened clubs, cited by name, in various cities in the U.S. Now there's a passel of information no reader of *Alice* should be without. I know I am richer for it.

There was some sprawling Martin Gardner did not do: He avoided areas which were nasty or heavy-handed. He almost never really interpreted textual material, though he pointed the way to prominent people who did—even Freud. He also avoided parts of the texts which might make readers uncomfortable—Carroll's ghastly class snobbery, for instance. Even when dark and itchy interpretation might seem inevitable, Gardner's regard for his readers was so great he abstained. Alice might cite poor Mabel in a snooty way, but Gardner is not going to call our attention to it. Nor should he in my view: I don't want to emerge from reading these great works with an aversion to their unlikeable author.

My favorite example of Martin Gardner's forbearance and graciousness comes at the end of Carroll's brilliant parody of Isaac Watts's "How doth the little busy bee." Most sneaky and even scholarly annotators (me included, Donald Gray too)[‡] go off on Darwin here. Spilling out bile on nihilistic darkness and Alice's insensitivity. Gardner, however, says only, "Carroll has chosen the lazy, slow-moving crocodile as a creature far removed from rapid-flying, ever-busy bee."That's it—and that's Martin Gardner: gracious, learned, and kind. We'll never look upon his like again.

‡ Gray's notes to *The Norton Critical Edition* of the *Alice* books, 1971, reprinted and updated frequently.

All in a Golden Afternoon

CHARLIE LOVETT

It's hard to believe that it's been over twenty-five years since I first began corresponding with Martin Gardner. It was four years later that we met for the first time. I was living in Winston-Salem running an antiquarian bookshop with my first wife, Stephanie; Martin was living in Hendersonville, about three hours' drive away, with his wife, Charlotte, an antiques enthusiast. In the spring of 1989 the Gardners came to town to attend an antique show and stopped by our house for a short visit. Although we corresponded for years, it was one of only a few times Martin and I met. It was a brief visit during which we exchanged a few books, but Martin invited us to come visit him at his home sometime, and in July of that year I spent one of the most delightful afternoons of my life talking about books, Carroll, and Carrollians with Martin Gardner.

He welcomed us warmly and was soon regaling us with stories and magic tricks. At one point it flashed across my mind that this must have been

Charlie Lovett is a former president of the LCSNA, a Lewis Carroll collector and scholar. He is the author of 13 books, including Lewis Carroll's Alice, Alice on Stage, Lewis Carroll and the Press, Lewis Carroll's England, *and* Lewis Carroll Among His Books. *His articles on Carroll have appeared in* Knight Letter, The Carrollian, Princeton University Library Chronicle, *and elsewhere. He has lectured on Carroll in the U.S. and abroad, including talks at Harvard University, the Smithsonian Institution, Oxford University, and at the first and second international Lewis Carroll conferences. He is currently the editor of* The Lewis Carroll Review.

what it was like to be in the presence of Lewis Carroll. It was in our early days (or so it seems now) of collecting and as we walked around his home, Martin would pull pamphlets and books off the shelf and ask, "Do you have this?" The answer was almost always, "No," and he would add the item to what became an armload of Carrolliana that he sent us home with, refusing all attempts at compensation. I remember in particular his copy, inscribed, of Stan Marx's rare publication of "Jabberwocky" in the Shaw Alphabet. We had recently purchased Stan's collection, and even he didn't have a copy of this item. To this day I've never seen another one.

Once we had toured the house and the bookshelves we settled down with a cup of tea and Martin showed us his latest project, what was soon to become *More Annotated Alice*. He had, he said, spent years collecting clippings, articles, and especially letters from readers, which provided fodder for new annotations of the *Alice* books. We went through the entire manuscript, chatting about the annotations and their sources and brainstorming about other possible notes. It was an amazing experience to be in the presence of this legend, whose work I had used many times, and who was treating me not as a "fan" but as an equal.

When we returned home, I typed up three pages of possible annotations, based on ideas I had had during our conversation and on a subsequent reading of the *Alice* books. I sent this list to Martin (I should add here that all our correspondence was typed letters sent through that most old fashioned of carriers, the U.S. Postal Service) and a few days later he replied:

> I must say, you sent a goldmine of ideas. A few suggestions I had already taken care of, but most of them were either totally new to me or had slipped my notice as possible spots for new notes. I have added numerous notes based on your memo, and duly credited them.

And so, because Martin's charm, excitement, and hospitality spurred my own thinking about the *Alice* texts, I was able to be a contributor to *More Annotated Alice*, and he was, of course as always, more than generous in his attributions.

We corresponded frequently over the next couple of years. Martin often needed a book we could provide in his preparations for *More Annotated Alice* (we provided one of the Peter Newell books from which the illustrations were taken, for instance). His meticulous attention to detail as he prepared the book for the press was a pleasure to watch as he described in his letters the ferreting out of various bits of information and allowed us to offer assistance in cases where a Lewis Carroll collector or antiquarian bookseller might be a valuable resource.

Following the publication of *More Annotated Alice* I continued to correspond with Martin, mostly on bookish matters. He asked me to help him track down volumes of *John Martin's Book*; I floated the idea of a book of parodies of "Jabberwocky" (on which, years later, Hilda Bohem did a more thorough job than I could have dreamed of doing).

When I left the antiquarian book business and we both left the state of North Carolina, Martin and I drifted apart and our correspondence petered out, but I shall always remember the excitement which accompanied the arrival of his letters, and I shall especially cherish the memory of the golden afternoon spent in his home. It was a day that I marked with a white stone.

Gardner in Japan

YOSHIYUKI MOMMA

From my point of view, if it had not been for *The Annotated Alice*, the *Alice* stories would not be as popular as they are in Japan now. Today at Amazon.co.jp, we can find at least 60 different current (not second-hand) editions of the *Alice* books in Japanese, including *Under Ground*, *The Nursery Alice*, and the Disney *Alices*. To my knowledge, in 2009 alone there were at least ten Alice-related exhibitions in Japan, although many of them were small-scale. *Pictures and Conversations: Lewis Carroll and Comics* has a section of "Japanese Comics Inspired by the *Alice* books," in which 28 Japanese comics (*manga*) are mentioned; that was five years ago and the number is steadily increasing. Alice is studied, from literature to comics, in many academic fields in Japan.

The Lewis Carroll Society of Japan was founded in 1994, and currently has around 150 members, some overseas. We hold meetings four times a year, and also have a general meeting, a Tea Party in July, and a Christmas

A serious Carroll collector, Yoshiyuki Momma is a founding member and the first chairman of the LCSJapan (1994), participated in the First (1989) and Second (1994) International Lewis Carroll Conferences, and has talked extensively about Alice in Japan. Besides the LCSJ, he is also a member and frequent attendee of the LCS(UK), the LCSNA, the LCSCanada, and the LCSBrazil. He and his wife, Naoko, translated Gardner's The Universe in a Handkerchief *into Japanese.*

Party. We publish *The Looking-Glass Letter* newsletter quarterly, and the *Mischmasch* journal yearly. But can we Japanese *understand* the *Alice* stories like those whose native language is English?

Martin Gardner writes in his preface:

> In the case of Alice we are dealing with a very curious, complicated kind of nonsense, written for British readers of another century, and we need to know a great many things that are not part of the text if we wish to capture its full wit and flavor. It is even worse than that, for some of Carroll's jokes could be understood only by residents of Oxford, and other jokes, still more private, could be understood only by the lovely daughters of Dean Liddell.

He also points out that "...the time is past when a child under fifteen, even in England, can read *Alice* with the same delight as gained from, say, *The Wind in the Willows* or *The Wizard of Oz*. Children today are bewildered and sometimes frightened by the nightmarish atmosphere of Alice's dreams."

Half a century has passed since he wrote this preface. Circumstances have changed dramatically these past fifty years: it is much more difficult today even for English children to understand the *Alice* stories. Is it even possible to understand the *Alice* books for those whose mother tongue is not English and whose culture is so very different from that of Victorian England? How can we Japanese capture their full wit and flavor? Fortunately, thanks to *The Annotated Alice*, we are able to enjoy them. I am sure that almost all scholars and students of Lewis Carroll in Japan, as elsewhere, have owed much of their understanding to *The Annotated Alice*.

When I was a university student, more than thirty years ago, I took a course in *Alice's Adventures in Wonderland*; I was immediately enchanted with it. I read the story line by line very carefully. With the help of *The Annotated Alice*, I realize how the story was made and how the quaint events were hammered out on the rowing expedition. But without the annotations, I could not have understood, for example, who the Duck, Dodo, Lory, and Eaglet were in real life.

Gardner's subsequent guidebooks have also enabled us to understand the stories more deeply. *The Annotated Alice* was translated into Japanese in 1980 in two volumes; *More Annotated Alice* was translated, also in two volumes, in 1994. These books further helped those who are not good at English to capture the wit and flavor of his classic works.

Beside the *Alice* books, many of Martin Gardner's books have been translated into Japanese. Not only Carrollians but also mathematicians, scientists, and general readers enjoy his books all over Japan. I was very fortunate to have had the chance to translate his *The Universe in a Handkerchief: Lewis Carroll's Mathematical Recreations, Games, Puzzles, and Word Plays* (New York: Copernicus, 1996). My wife, Naoko Momma, and I did our best to translate it into Japanese, hurrying to get it published two years later (Tokyo: Hakuyosha, 1998), the centennial of Carroll's death.

Alan Tannenbaum emailed me on December 7, 2005 that Martin Gardner showed him a copy of the Japanese *Handkerchief* when he and Alison visited him at his new home in Oklahoma. I am very glad that Martin Gardner had a copy of our translated book on his bookshelf. I hope he liked it and knew how much we Japanese love Lewis Carroll and his works, and his own, too.

Martin Gardner & Lewis Carroll: The Magical Connection

CHRISTOPHER MORGAN

I first met Martin Gardner in 1981, at his home in Hastings-on-Hudson, New York. I had come to invite him to write some articles for *BYTE* magazine (an early personal computer magazine of which I was Senior Editor). I wanted him to discuss the connection between recreational mathematics and computers. He ultimately declined, because he felt he didn't know enough about computers. Since we were both amateur magicians (Martin was well respected in the magic community as the inventor of scores of clever tricks, many mathematically based), our meeting quickly turned into a magic session. We traded tricks for several hours. For me it was, as Lewis Carroll would often say when recording memorable days in his diary, a "white stone" day.

As I got to know Martin better over the years, I was struck by how much he reminded me of Lewis Carroll, the subject of his most popular book, *The Annotated Alice*. Both shared a common love of logic, recreational mathematics, chess, puzzles, games, cryptology, and magic. But they also

Christopher Morgan was editor-in-chief of Byte *magazine, the founding editor of* Lotus Magazine and Popular Computing, *and was editor-in-chief of McGraw-Hill's* Byte Books *series. He has also written* The Official Computer Bowl Trivia Book, *and* Wizards and Their Wonders: Portraits In Computing. *In his spare time he is a musician, audiophile, book collector (Carroll is a specialty), president of the Ticknor Society (bibliophiles), photographer, and an organizer of the Gathering4Gardner and the Computer Bowl.*

shared many personality traits: a sense of playfulness, an appreciation of whimsy, a love of the absurd, a delight in logical fallacies, and—perhaps most tellingly—a joy in learning new things. These are many of the same traits I see in the magicians I know. I think the common magical connection is no accident. It helps to explain why Martin was so drawn to the works of Carroll.

Martin's seminal book about mathematics and magic, *Mathematics, Magic, and Mystery* (Dover, 1956), could very well have been written by Carroll. They both loved to conceal secrets, whether through Carroll's use of acrostics in poems to hide people's names, or the clever numerological feints constructed by Martin's alter ego, Dr. Matrix. Martin once said that "Mathematical magic, like chess, has its own curious charms. [It] combines the beauty of mathematical structure with the entertainment value of a trick." Both Carroll and Gardner liked to show tricks to small groups of friends, but neither performed for large groups on stage. Carroll restricted his magical presentations mostly to his child friends and family. Martin had an aversion to performing for groups, but enjoyed entertaining friends at home.

In his wonderful collection of Carroll's games and puzzles, *The Magic of Lewis Carroll* (Simon & Schuster, 1973), John Fisher says:

> The itch to baffle and to mystify, and to entertain into the bargain, remained with [Carroll] until his death. . . . With a magician's instinct for tracking down the impossible, he was able to apply something more than the straightforward academic approach to his studies in mathematics and logic, sources of mystification no conventional magician had ever tapped. . . . Not only did mathematics present an infinite means of bewilderment, in stark contrast to the stereotyped repertoire of the drawing-room entertainer, but even more remarkably often displayed a baffling aspect which he, the magician himself, could not account for. Logic professed fallacies which, when decked out with entertaining characters, were far more spellbinding than the contrived red-herring patter used to misdirect an audience's awareness away from a more tangible secret. Here was as near as he would ever come to discovering real magic.

Tellingly, this passage could have been written about Martin Gardner as well as Carroll. Martin's joy in learning new things never left him. He was ninety-one years old when I visited him in 2006, but he seemed more like a twenty-year-old in his eagerness to jot down the details of a magic trick that was new to him. Carroll's diaries are full of passages recording new discoveries, even late in his life. For example, his diary entry for September 24, 1891, when he was fifty-nine years old, says:

> An inventive day. It has long been a 'desideratum' with me to be able to make short memoranda in the dark, without the unpleasant necessity of having to get up and strike a light. . . .Today I conceived the idea of having squares, cut out in card, and devising an alphabet, of which each letter could be made of lines along the edges of the squares, and dots at the corners. I invented the alphabet, and made the grating of sixteen squares. It works well. I shall call it 'The Typhlograph.'*

Another striking common ground between Gardner and Carroll is their love of humor, and their use of it to explain and delight. The *Alice* books are full of humor, and Gardner's books are full of wit, funny poems, and practical jokes. (Martin loved to recite funny limericks—some decidedly not G-rated!)

The link between humor and creative puzzle-solving may be more than just coincidental. Recent research by Mark Beeman, Associate Professor of Psychology in the Cognitive Neuroscience Program at Northwestern University, shows that one's mood can affect one's ability to solve puzzles.† Specifically, people who watched humorous videos before attempting to solve puzzles did better than a control group who didn't watch the videos. Our brains seem to become more creative and open to new ideas when we're in a good mood.

* Carroll changed its name to the "Nyctograph" a month later. The Nyctograph is discussed in *Knight Letter* 75:8–10 and was the subject of a pamphlet, *Square Alice*, privately printed by Alan and Alison Tannenbaum and distributed to members of the LCSNA in attendance at the Spring 2005 meeting. – Ed.

† Subramaniam, K., Kounios, J., Parrish, T. B. & Jung-Beeman, M. (2009). A brain mechanism for facilitation of insight by positive affect. *Journal of Cognitive Neuroscience* 21, 415–432. Reported in "Tracing the Spark of Creative Problem-Solving" by Benedict Carey in *The New York Times* Science Section, December 7, 2010.

Martin Gardner would probably not have been surprised by this, since he spent his entire life popularizing the notion of "recreational mathematics." Gardner and Carroll both seemed to realize that we learn best when we're in a state of delight and wonderment—a "magical" state. Wherever the two of them are today, I suspect they're having the last laugh.

Gardner in the U.K.

MARK RICHARDS

Soon after the Lewis Carroll Society (UK) was formed in 1969 it became clear that there was interest in such a society around the world and that many of the best known writers who had published works about Lewis Carroll were keen to take part.

Martin Gardner was one of those early members whose participation gave the Society a certain credibility. His *Annotated Alice* and *Annotated Snark* had demonstrated to many the depth of Carroll's work, which led to their joining the Society. To know that a writer and scholar of Gardner's importance was a member—and an active one, at that—made being a member something rather special. There was, and still is, a thrill to being in the company of such experts.

I was a follower of Gardner's work before I became interested in Carroll and, like so many others, I can trace the emergence of mathematics as my "best subject" at school, and subsequently as the discipline of my degree,

Mark Richards has been studying the life and works of Charles Dodgson for 35 years in what he describes as a "dilettante fashion," although he has made a contribution to the understanding of Carroll's unpublished logic papers. He is an omnivorous collector of Carroll-related printed matter and ephemera, with a particular interest in illustration and fine art. Richards has been a committee member of the Lewis Carroll Society (UK), on and off, since the 1980s and is currently Chairman of the Society and Executive Editor of The Carrollian.

back to the period when I avidly read his "Mathematical Games" articles in *Scientific American*.

Gardner remained a loyal member of the LCS right up to his death and I, like many other officers of the Society, have enjoyed corresponding with him and we have all been grateful for his continued support spanning those forty years.

The last letter we received from Martin was dated May 6, 2010—little more than two weeks before he passed away. A couple of weeks earlier he had received some correspondence regarding the abandoned screenplay that Aldous Huxley had written for Walt Disney, which combined an adaptation of *Alice* with a biography of Dodgson. Martin's letter to me reads:

> Dear Mark:
> Herewith a copy of a letter from a friend. Aldous Huxley's film script was news to me! If you think it news to most Carrollians, it might be worth mentioning in The Carrollian. I have no idea if the script is mentioned in any books by or about Huxley.
> Best,
> Martin

There had been an article about Huxley's script in *Knight Letter* ('Aldous in Wonderland,' *KL* 49:6) some ten years earlier and references in subsequent issues. We can certainly understand that an elderly man, with so many wide-ranging interests, might have overlooked or forgotten about that article, but the interesting point is that right up to the last few weeks of his life, he continued to have an active interest in Carroll, to the point where people were still sending him information. And, to our delight, he continued to do his best to share this information and support our societies.

As officers of the Lewis Carroll societies, whether it is in the U.K., U.S.A., or elsewhere, we devote much of our time to promoting interest in the life and works of Lewis Carroll. In this venture, none of us can achieve a fraction of what Gardner did with his two annotated texts, but it has been easier and more pleasurable doing what we can knowing we had his support and help when we needed it.

Memories of Martin

DAVID SCHAEFER

I met Martin Gardner and his wife at Princeton University in 1974 at the formation meeting of the Lewis Carroll Society of North America. We had many pleasant interactions in the years that followed, which included discussions about the "Wasp in the Wig" and about the illustrations for his *More Annotated Alice* (1990).

Martin was planning to use Peter Newell's from 1901/02; we speculated about whether they had ever been seen since the original Harper & Brothers editions. In the course of our discussion I remembered that in our collection was a tome published in 1967 entitled *Alice in Womanland, Or the Feminine Mistake*, which had used several. So *More Annotated Alice* would be the second to make use of them.

Martin always encouraged my interest in Lewis Carroll–related motion pictures. He honored me by including "Alice on the Screen," a listing of films I compiled for him, in the Definitive Edition of his *Annotated Alice*.

Martin was a great guy and I feel honored to have known him.

David Schaefer is a former president of the LCSNA and the premier scholar and collector of Alice-related films.

Martin Gardner: A Personal Reminiscence

JUSTIN G. SCHILLER

Martin Gardner and I first met on January16, 1956, an especially auspicious date, as it was the official opening of Columbia University's centenary exhibition for L. Frank Baum. I was twelve years old at the time and became the youngest lender to Columbia's library in its two hundred year history, since several items from my personal Baum and Oz collection had been included in this exhibition, detailed by an Addenda at the end of the published catalogue.

As I had met Oz-author Jack Snow the year before, my parents and I invited him to join us in our automobile as we traveled to Butler Library on Manhattan's Upper West Side. When entering the building, Martin was descending the staircase and greeted Jack. One year before, Martin had serialized a scholarly introduction to Baum in the January/February 1955 issues of *Magazine of Fantasy and Science Fiction,* afterwards revised as "The

Justin G. Schiller is a principal founder of both The International Wizard of Oz Club in 1957 and the LCSNA in 1974. A book collector by avocation and a rare book dealer by trade, he specializes in antiquarian children's books and related illustration art. He has written extensively on the subject, including Realms of Childhood *(1983),* Nonsensus *(1988),* Digging For Treasure: An Adventure In Appraising Rare and Collectible Children's Books *(1997), and* Alice's Adventures in Wonderland: An 1865 Printing Re-Described and Newly Identified as the Publisher's "File Copy" *(1990), which is both a bibliographic census of the 1865 printing and a description of the copy that he sold for $1.54 million in December 1998, still the highest price ever paid for a children's book.*

Royal Historian of Oz," the introduction to his *The Wizard of Oz and Who He Was* (Michigan State University Press, 1957), which perhaps was how they knew one another. And Jack's copy of the original poster advertising *The Wonderful Wizard of Oz* (Chicago: Geo M Hill, 1900) was reproduced from his personal collection on the front cover of this later version of Martin's essay. Jack introduced me and my parents to Martin, and I remember how impressed I was when Martin said "Ah, you're the young man I read about" referring to the single exhibition case upstairs that contained some modest items from my collection, compared to the great treasures that surrounded them. His graciousness and encouragement were immediately obvious, but I don't think we exchanged addresses nor did anyone anticipate how soon we would meet again.

On July 13, 1956, Jack Snow died, and Martin became his executor on behalf of the Snow family in Ohio. Using Jack's address book Martin telephoned our house with the sad news: there would be no local funeral, and Martin was in charge of selling off what remained from Jack's collection. There must have been a brief descriptive list sent via the mail, so we made an appointment to visit Martin at Jack's apartment, but there was not much extra money to spend: my father was a traveling salesman for a high-end clothing line, which physically required him visiting men's shops along the East Coast with sample cases showing formal and dress shirts, while my mother was just beginning her own business in antique jewelry. I was still one year away from entering high school.

During the year plus that I knew Jack Snow, I would telephone him at least once a week and he always seemed glad to hear from me, but he often complained about the quantities of mail he received regularly from Oz enthusiasts everywhere—many of them asking him questions about Oz, or sending him new stories of their own creation, and always requiring some response. Jack mentioned he wished they would create their own club and write to each other without involving him in the equation. So I mentioned that to Martin and asked if he might have addresses of those people who wrote Jack, as we could write to each of them and create an Oz club in Jack's memory. Martin thought that was a good idea, but he had just sold all of this correspondence to a school teacher in Escanaba, Michigan. On my behalf he offered to write Fred Meyer to inquire if he could share these names and addresses of Oz people with me. And so began the International

Wizard of Oz Club in January 1957, with Martin, Fred, and Frank Joslyn Baum being honorary officers, while my parents and I initiated sending letters of invitation to all of Jack's correspondents.

As I grew older in Oz and attended university, Martin Gardner and his wife Charlotte were always generous with their time and advice. I can remember visiting them on several occasions in Hastings-on-Hudson during my adolescent years and being treated with the maturity I hoped to have one day rather than as someone living in a fantasy world (which was probably closer to reality). Although Oz first brought us together, I was perhaps most influenced by Martin's work with the texts of Lewis Carroll. His *Annotated Alice* (1960) and *Annotated Snark* (1962) were role models for me. I read each of them several times and came to realize the importance of understanding not only what the actual texts cryptically reveal but also the importance of background history behind their creations. They were partly my inspiration for becoming a serious researcher, and delving into Dodgson's writings became palatable due to my friendship with Martin and the ease with which he unraveled their mysteries. And when I sold my Oz library at Swann Auction Galleries in November 1978, Martin contributed a page-and-a-half foreword reminiscing about Oz as an American Fairyland, and our friendship over the past twenty-plus years.

Because I was exploring many directions in searching around the Fourth Ave. secondhand bookshops of New York City, I uncovered unusual paper curiosities and can remember Martin showing me the ephemeral "Vanishing Chinaman" rotating game card of Sam Lloyd—or perhaps I found an example and discovered the work Martin had done on it. Somehow I was never surprised that he seemed to know about *everything* but handled this knowledge with matter-of-fact humility. And he was always willing and open to share information with anyone who came to him with a legitimate purpose in mind.

At one point he heard of my own interest in collecting "Nieuwe Kunst," both the decorative and fine art of the Netherlands dating from the 1880s through the 1920s, what might otherwise be classified as Dutch Art Nouveau and Art Deco. My partner and I had purchased a rambling old house in New York State's lower Hudson Valley, which was an old Dutch settlement going back three hundred fifty years. We were in Amsterdam to attend an antiquarian book fair in March 1986 and Sotheby's was concurrently exhibiting and then selling off two lifetime collections. We were already under contract to buy our home, but closing would not be until the end of June so the children of the owners could finish their academic semester. Never would I imagine that such art would filter also into Martin's world, but it did—from his friendship with M. C. Escher, whom he publicized in a 1961 article on Escher's visual optical illusions for *Scientific American*.

Although I always felt Martin was accessible to my inquiries, once he and Charlotte moved out of New York (in 1979) we seldom communicated except by way of sending each other occasional publications that would overlap mutual interests. He was always a kind, generous critic, with encouraging words and occasionally a well-constructed suggestion.

Today, it seems amazing that it has already been a dozen years since his own Oz chronicles were published: *Visitors From Oz: The Wild Adventures of Dorothy, the Scarecrow and the Tin Woodman* (New York: St Martins Press, 1998). Meanwhile I had been busy running in many directions, keeping two antiquarian rare books businesses active, so my lack of contact with Martin made me feel somewhat shy in writing regularly. But I can still hear his voice in my mind from more than a quarter-century ago, and I always knew that if I needed his help or advice, he would be receptive as ever, just as if we had been constantly in touch.

One might think of Martin as a Renaissance man from all the different "hats" he wore, but the multiple facets fitted together neatly as a jigsaw. He was certainly a pioneer on many fronts, and unique for his creative genius in seeing new twists in the world around us. And he was certainly an icon of his generation, a legendary name known to many, always associated with the highest standards of integrity and scholarship. But most importantly to me, he was a dear friend. I'm sad only in not having expressed my appreciation more openly in recent times, nor to have had the foresight and intelligence to say goodbye.

A Remembrance of a Gracious Carollian

BYRON SEWELL

I am one of those people who never read *Alice's Adventures in Wonderland* as a child. However, I am old enough to have seen and remember the original release of Disney's cinema version in 1951, so I at least knew a little about the story. My first real encounter with *Alice* was in 1970 when I spotted Martin Gardner's *The Annotated Alice* in a bookstore and on a whim purchased a copy for my first wife.

As it happened, a few months later I was assigned to work in London for a year. We took the book along, deciding that while we were in London it would be fun to buy some old *Alices*. We eventually had considerable success in starting a collection (that is now housed at The Harry Ransom Humanities Research Center at the University of Texas at Austin). That year I also discovered *The Hunting of the Snark* and I decided that I should do an illustrated version of my own. One thing led to another and I finally found a

Byron Sewell is a noted Carroll collector, author, bibliographer, parodist, and illustrator. His book illustrations include The Hunting of the Snark *(Catalpa Press, 1974),* Alitji in the Dreamtime *(Pitjantatjara translation, 1975), and* An, Sun-hee's Adventures under the Land of Morning Calm *(Korean translation, 1990); bibliographies* Much of a Muchness *(American editions of Alice, 1992),* Pictures and Conversations *(Carroll in the comics, 2003),* An Annotated Bibliography of Lewis Carroll's Sylvie and Bruno Books *(2008); his Carrollian parodies, short stories, and ephemera are too numerous to even begin to list here.*

publisher in Ian Hodgkins, who owned the Catalpa Press. Mr. Hodgkins and I decided that what we needed for the book was an introduction by some famous Carrollian. I had not yet discovered the Lewis Carroll Society and with my limited knowledge about collecting Lewis Carroll the only name that came to my mind was Martin Gardner.

Having nothing to lose I decided to write and ask him if he would pen the introduction for my *Snark*. I included a set of poor quality black and white snapshots of the illustrations in my letter so that he could see what the book might look like. I explained that I could probably not afford to pay him his normal fee for writing such a thing, but said that I would send him a copy of the book and one of the original illustrations as a token thank you. A few weeks later, to my great delight, I received a gracious letter from him informing me that he agreed to do it. His letter was enthusiastic about the illustrations and he even had a few suggestions about some special effects that I should consider. A short time later the draft of his introduction appeared and I was astonished at the wonderful things he had to say. After numerous setbacks and problems the book was finally published in 1974. At Gardner's suggestion the final illustration was printed in black with a red Baker, along with a sheet of red film that would make the Baker vanish when placed on top of the illustration. I sent the illustration of Lewis Carroll sitting in a rowboat that appears at the end of the book to him, along with a copy of the edition.

In a letter of March 21, 2000 Martin wrote to me that "The original drawing that you sent me from your *Snark* is nicely framed and hanging in my library." This same letter includes the following interesting comments on other subjects:

> The new Definitive Edition of Annotated Alice sold out entirely through the big houses on the Internet, and is now in a second printing. Penguin will be issuing a paperback edition later this year or in 2001. My Oz fantasy, Visitors from Oz, published last year, fell on its face. One reviewer, for the Washington Post, called it a "poor thing of a novel." Then to my surprise, the Penguin paperback got a full-page favorable review in the London Times! I find it curious that L. Frank Baum, America's Lewis Carroll, seems to be more appreciated now in England than here!

Over the past four decades or so I have produced a rather large number of books, pamphlets, short stories, parodies, and bibliographies, a few paintings, lots of illustrations, and various bits and pieces of ephemera, all on a Carrollian theme. In a very real sense, for good or bad, all of this "stuff" is the direct result of Martin Gardner having introduced me to the *Alice* books through his marvelous *The Annotated Alice*. He must share the blame or praise, for had he never written his famous book none of my own things would have ever appeared.

Annotations of Immortality

BRIAN SIBLEY

It had been my favorite book for as long as I could remember. Even before I could read, I was plaguing any willing family member with demands that they read (or *re*-read) me Lewis Carroll's *Alice's Adventures in Wonderland*, until I knew it so well that I could quote whole passages by heart.

In the process of growing up, I remained devotedly loyal in my affection for the extraordinary denizens of Wonderland and (when I later discovered them) Looking-glass World and Snark Island, but as other authors and books increasingly elbowed their way into my consciousness, Alice & Co., might easily have been crowded out of my affections for ever. Then, when I was eleven years old (and, not untypically for a British schoolboy, was being urged to "put aside childish things"), I had the great good fortune to stumble on Martin Gardner's *The Annotated Alice*.

The experience was akin to Alice finding the little gold key on the glass table in the Hall of Doors. Unlike Alice's frustrating experience (either the

Brian Sibley is a former Secretary of the Lewis Carroll Society (UK) and was the founding editor of its newsletter, Bandersnatch. *He has written and broadcast extensively on fantasy literature and films and has created award-winning dramatizations for radio and stage of the works of, among others, J. R. R. Tolkien, C. S. Lewis, Edward Lear, Charles Dickens, and Mervyn Peake. The author, co-author, or editor of more than twenty books, he is also a passionate devotee of magic, and currently serves as Chairman of Council of the renowned, century-old The Magic Circle in London.*

locks being too large or the key being too small), Mr. Gardner's key opened door after door, revealing that—despite my parrot-like aptitude—I had scarcely more than a superficial understanding of texts that I thought I knew so well.

True, I was annoyingly smug whenever I came across an annotation that was clearly intended to explain the peculiarly eccentric habits of the English to an American readership, but that didn't negate the rich pleasure gained from an increasing appreciation of Lewis Carroll's puns, parodies, puzzles, and paradoxes or the discovery of the historical, social, and literary references with which the works are sprinkled as liberally as the Queen's tarts were with pepper.

Most importantly, Martin Gardner introduced me to the complex personality that was Charles Lutwidge Dodgson. Reading—more accurately, *consuming*—*The Annotated Alice* was the beginning of what has been a lifelong fascination with the life and work of the Tweedledee and Tweedledum personas of Mr. Carroll and Mr. Dodgson.

And while Gardner sent me plunging down any number of new rabbit holes in pursuit of further Carrollian biographical and literary analysis, he remained a frequent and fond companion, not just in wanderings through Wonderland, but on other wild adventures, too: riding the tornado with Dorothy to Oz, taking ship with the Ancient Mariner, or infiltrating the anarchic Council of Days with the Man who was Thursday.

I may have been out of my depth in the deeper waters of science and mathematics plumbed by Gardner, but I shared his love of logic, his passion for puzzles, his fascination with pseudoscience, philosophy, and religion, and his enthrallment with magic and illusion. Indeed, it was Gardner's writings on legerdemain, along with his *Alice* annotations linking Lewis Carroll with such magical paraphernalia as top hats and white rabbits, that fired and fueled my own curiosity about the conjuror's craft.

My partner, the magician David Weeks, also has particular cause to be grateful to Gardner, the trickster. One evening, in a small village at the very end of a small Greek island, a biker—looking like Otto von Bismarck in a remake of *Easy Rider*—arrived at the small waterside taverna where David was performing magic for fellow holidaymakers. At the end of the routine, the biker challenged the conjuror to solve a matchstick puzzle.

The skeptic was throwing down a gauntlet. If the magic-man failed to rise to the challenge and solve the conundrum, his most dexterous sleight of hand would count for naught. The answer, David was confident, would be in his holiday reading—Martin Gardner's *Encyclopedia of Impromptu Magic*. Slipping away to his room, he returned, minutes later, not just with the solution to the poser, but with three even more bamboozling puzzles with which to stymie his opponent!

"Life, what is it but a dream?" asks Lewis Carroll at the end of his acrostic poem to Alice Liddell. Magic and dreaming have much in common: the laws of the universe and the perplexing rules and regulations of our humdrum existence can be turned upside down and inside out so that they can be laughed at or puzzled over or re-envisaged through the fanciful freedom of the human imagination.

Lewis Carroll was a trailblazing explorer in the realm of topsy-turvydom and we are fortunate, beyond measure, to have had so enquiring and diligent a scholar as Martin Gardner to have applied his mind to unraveling the riddles and elucidating the enigmas of Carroll's dream-worlds with the precision of a mathematician, the commonsense of a logician, and the flair and showmanship of a magician.

In the Garden of the Snark

MAHENDRA SINGH

Martin Gardner's *The Annotated Hunting of the Snark* was my first reading of that poem. I was a young lad and I confess that Gardner's annotations fascinated me as much as Lewis Carroll's masterpiece. I had already waded through a few academic analyses of other literary works that interested me and although they were useful, they simply were not fun. Nor were they so catholic in their pell-mell tastes, and I think that it was this intellectual exuberance of Gardner's that first lured me into the arcane hunting grounds of Carroll's nonsense epic. When one considers his methods and outlook, Martin Gardner was perfectly suited for hunting Snark and for persuading others to join him in the chase.

Gardner was possessed by the Victorian mania for collecting and disseminating facts: wonderful, paradoxical, and occasionally goofy facts. He unfolded the *Snark*'s 715 line, anapestic origami into a flattened description

Mahendra Singh lives in Montreal, where he has gained a reputation as an illustrator of all things Lewis Carroll and Surrealist. His teenaged exposure to both Martin Gardner's Annotated Snark *and Max Ernst's* A Week of Kindness *were decisive influences in his artistic career. He is one of the editors of the* Knight Letter *and has illustrated a deeply witty, Surrealist, graphic-novel version of the* Snark, *published by Melville House in 2010. Singh's blog at justtheplaceforasnark.blogspot.com is quite amusing and informative.*

of the world as Carroll had found it, and then let us romp along with him from one end of the century to the other. The triunary mathematical obsessions of semi-obscure Edwardian philosophers, the oral fixations of British gentlemen going walkabout amongst the hoi polloi, the Snarkish table talk of President Theodore Roosevelt and Edith Wharton: Gardner's itinerary made room for all of them and much more.

At the same time, his annotations were also a sly poke at the Victorian tradition of self-improvement. The dissemination of useful knowledge through public lectures was the classic modus operandi of the Victorian scientific popularizer, and Carroll himself had deftly parodied this popular lecturing mania in the Beaver's Lesson and the Bellman's Five Marks of the Snark. Gardner made the *Annotated Snark* into a Carrollian lecturer's indispensable vade mecum—assuming that one wished to lecture the masses upon the utility of genuine Nonsense, of course!

This is not to say that Gardner was merely a nostalgist. He did not eschew chasing his Snark out of the nineteenth century and into the twentieth century thickets of postmodernism. There was a delightful frisson of circular ruin–ation in his annotating of what was essentially an annotation of reality itself, in his Borgesian catalog of the *Snark's* many intersections of Victorian reality and the Carrollian Multiverse. The poem itself, which occupied the Nonsensical heart of darkness within that Multiverse, referred to many things, some real and some not, but it rarely provided any logical connections on its own. It was a code with no message, a *simulation of the appearance of meaning* designed to baffle even the most obsessive Victorian taxonomist and popularizer.

Gardner deftly sidestepped this philosophical pitfall. His *Annotated* books generously entertained every explanatory theory possible and even some that were highly improbable. His own attempt at the Holy Grail of Snarkian research, the quest for meaning, was relegated to a Preface in which he came down on the side of existential dread, perhaps more for the sake of providing elbow room for the idea of redemption than anything else. Gardner may have been only an amateur mathematician but he understood that existentialism was not a philosophy at all, only faith nibbling at logic's edges. It was an sincere compromise between faith and logic on Gardner's part, and the religious-minded logician Carroll probably would have approved of this, concerned as he had been with improving the happiness of Snark hunters, not just their minds.

In any case, the *Annotated Snark* provided an entertaining key to Carroll's simulated code. The key could not definitively unlock the door, but in this nonsense universe of simulated doors, it was the least irrational (and most aesthetically right) thing to do. In the final analysis, Gardner was a rationalist in the classical British tradition — stiff upper lip and never go native, especially when roughing it in the jungles of Nonsense!

Some readers might think that such a rationalist spirit would be anathema to the free spirited lunacy of Carrollian nonsense but no, it had

Lewis Carroll, from *The Hunting of the Snark*, illustrated by Mahendra Singh (courtesy of Melville House)

everything to do with it. Carroll's baroque variations upon logic and words were essentially rational daydreams, the idle speculations of a fiendishly logical mind with a bit too much time on its hands. In the hands of the practically inclined, such speculation becomes scientific theory; in the hands of the incorrigible dreamer, it becomes art. This was another of Gardner's larger achievements, his emphasis upon the ancient connection between art and science, a connection which the classically educated Carroll had utilized as one of the building blocks of the *Snark*.

This logical connection has fallen into some disuse now; the concepts involved are often considered too difficult for popular consumption in our age of hyper-specialization. It's not fun, as they say, so modern youth cannot cope with it. But in Gardner's hands, science was fun. Art was fun. Logic was cool. The pleasures of the mind were not despised as elitist luxuries nor dismissed with deprecating irony or populist contempt. They were all the happy fruits of a fertile intelligence, and Martin Gardner spent his life cultivating them for the enjoyment and benefit of others.

The *Annotated Snark* made it clear to younger readers that the somewhat jumbled contents of an educated person's mind, all one's fascinating logical and aesthetic manias, one's intellectual quirks and tics, all the books you had read and places you had been to, all of them were perfectly acceptable to intelligent adults, provided that they were logically valid and aesthetically sound.

It was pretty hot stuff for a young bookworm in the 1970s, and it still is. Gardner was tackling one of the great cultural dilemmas of our time. The beauty of logic and the logic of beauty have become passé, neglected if not openly ridiculed by those who should know better. The basic tenets of science are routinely ignored in public affairs and as for the arts, they have been relegated to a therapeutic ghetto. The *Snark* may be Nonsense but it cannot hold a candle-stub to the lunatic impulses shaping our twenty-first-century culture. Reading Gardner's erudite and humorous commentaries upon the *Snark* reassures young people that no matter how isolated they feel intellectually, they are not alone. Somewhere out there, there are still Carrollians who can read their *Snark* with their feet on the hearthrug near the fender, and then "explain all the while in a popular style / which the Beaver could well understand.

Reminiscences

DAVID SINGMASTER

I was an intermittent correspondent with Martin Gardner for about forty years, and had the pleasure of visiting him twice. Here I relate some miscellaneous episodes of this history.

Martin's first mathematical article was on Flexagons in the December 1956 issue of *Scientific American*. I was in my first year at CalTech and can still remember that soon thereafter nearly everyone was enthusiastically folding little strips of paper; this was my generation's introduction to the wonders of recreational mathematics, and to Gardner and his many other interests as well.

My first published paper was "On round pegs in square holes and square pegs in round holes," *Math. Mag.* 37:5 (November, 1964) 335–337. In this I examined the obvious ratios of the fit of a circle (or a sphere in n dimensions) to its circumscribed square (or n-dimensional cube) and of a square (n-dimensional cube) to its circumscribed circle (sphere in

American-born David Singmaster is a retired professor of mathematics at London South Bank University, England. A self-described metagrobologist (designer, maker, collector, and solver of puzzles), he is most famous for his championing of, notation for, and solution to the Rubik's cube, on which he has written or co-written several books, such as The Cube: Secrets, Stories and Solutions of the World's Best-selling Puzzle, *and for his huge personal collection of mechanical puzzles and books of brainteasers. He has contributed puzzle columns to newspapers, magazines, radio, and television shows, and has acted as Conference Chairman for four of the Gatherings for Gardner.*

n-dimensions) and found that the circle fits better in the square than the square in the circle, and that this held up through dimension 8. Gardner was intrigued by this and mentioned it in his column, "Packing: Circles and Spheres" in his May 1968 column, and subsequently in his 1979 collection *Mathematical Circus*.

Over the following years, we corresponded on a number of topics. In 1978, I encountered the Rubik's Cube and it took over my life for some time. In December 1979, I visited my family in New York and went to visit Martin at his home in Hastings-on-Hudson. I brought a Rubik's Cube for him—I recall he hadn't handled one before. He showed me around his house. There were 29 (or so) filing cabinets containing material relating to his columns. When I asked about a topic, he would walk over to a particular cabinet, open a particular drawer, and extract the relevant file. Back on the ground floor, I noticed an enclosed veranda full of books. I asked what these were and he said, rather dismissively, that they were foreign editions and translations of his books.

We then went down to the cellar where he kept various puzzles and toys that people were constantly sending him. He didn't really collect these, and said that Tom Ransom, a noted puzzle collector in Toronto, came down with a van every six months or so and took them away. Martin then opened a refrigerator, reached in, and handed me two puzzles. I asked if these were hot puzzles, but he said the fridge had long ceased to work, but it had lots of shelf space. One of these presents was an elegant model of a 3-dimensional projection of a 4-dimensional cube, made from thin metal strips with notches to let it hinge and fold flat. It has an identification—©1972 Eytan Kaufman N.Y. N.Y.—on it, but I've never seen or heard of another example. I still use it in lectures on polyhedra, and it has been used by British television personality Johnny Ball. (Eytan Kaufman is a now distinguished structural engineer in Manhattan, but I don't know if more examples have ever been produced.) The other present was *Solid Mechanism No. 16*, designed by Shiro Takahashi. This was like a Necker cube with joints allowing the front face to rotate relative to the back face, causing the apparent cube to move between the two appearances of the Necker cube. (Sadly, it has since come unsoldered. Takahashi has a website; this object is Work #6 and I am hoping to get another example.)

I visited Martin in Hendersonville in January, 1991. Here I saw the famous stand-up desk that he used for writing. He had found his house too small to accommodate all his material and had taken an additional apartment just to hold his books and papers, though he had weeded some material, and now only had about 20 filing cabinets. He had to go off to an antiques convention with Charlotte but kindly left me keys. I spent about two days reading through material, then photocopying a good deal of it at the local library. While talking with him, I discovered his interest in April Fools tricks. He said he was considering writing a book on them. (Although the book never got written, he was famous for his April Fools Day columns in *Scientific American*.) I, too, have a large file of these and sent him copies when I got back to London. He was delighted with them—he said he hadn't seen any of the U.K. examples I sent, not even The Great Spaghetti Harvest of 1957.

As readers may know, Martin was interested in a number of offbeat authors: G. K. Chesterton, L. Frank Baum (of "Oz" fame), Lord Dunsany, obscure authors of famous poems, etc. He was editing the works of G. K. Chesterton, but he said he'd never been to England and would like some first-hand description for *The Napoleon of Notting Hill*, which is set in Pump Street, Notting Hill. I told him that I was pretty knowledgeable about London and I would look up Pump Street. Getting back to London, I soon discovered that there is no such street and never was. I copied out various maps of the area and then went around it on foot, photographing street views, house frontages, etc. I got stopped by a security guard at one point—apparently Princes William and Harry were at school in one of the houses; the area is just a few blocks from Kensington Palace. Martin was so pleased to find that Chesterton had perpetrated a hoax that he wrote it up as Chesterton's Pump Street Hoax, which appeared in *Midwest Chesterton News* 3:8 (May 10, 1991), p. 4.

In 1993, Martin asked me to get copies of articles on Lewis Carroll's Barber Paradox which had appeared in *Mind* in 1894, 1895, and 1905—none of the libraries near him had copies going back that far, but it was easily done in London.

Though my relationship with Gardner was not as large as his was with many others, it was always a great pleasure to correspond with or talk to him and I feel honored to have helped him several times. Looking over the letters from him shows the curiosity, warmth, and humanity that will be familiar to anyone who knew him. We will not see his like again.

A Day in Norman

ALAN TANNENBAUM

In October, 2005 my wife, Alison, and I stopped in Norman, Oklahoma, to see Martin Gardner on our way from Austin to Des Moines to attend the fall meeting of the Lewis Carroll Society of North America.

I first read Martin's columns in *Scientific American* in the late '60s and he was responsible in large part for my interests in, among others, M. C. Escher and Lewis Carroll. He put me and my friends onto John Horton Conway's Game of Life, the first viral recreational programming challenge for computer geeks. At IBM in the mid-'70s, my group used a clever paper puzzle called a "hexaflexagon" to promote one of our products at a trade show; we had seen it on a TV show that featured Martin and so got the required permissions from him and CBS, and it was a wonderful success.

I had corresponded with Martin over the years about Lewis Carroll–related subjects, and now I was about to meet a man who, in my eyes, was a genius. When we met he turned out to be about the friendliest man you could imagine; very open and very gracious.

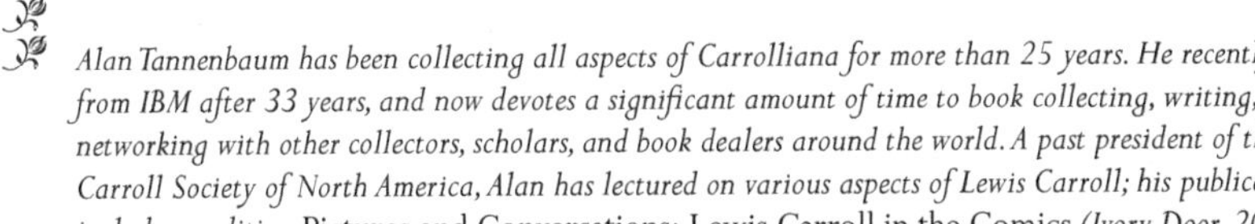

Alan Tannenbaum has been collecting all aspects of Carrolliana for more than 25 years. He recently retired from IBM after 33 years, and now devotes a significant amount of time to book collecting, writing, and networking with other collectors, scholars, and book dealers around the world. A past president of the Lewis Carroll Society of North America, Alan has lectured on various aspects of Lewis Carroll; his publications include co-editing Pictures and Conversations: Lewis Carroll in the Comics *(Ivory Door, 2003) and Square Alice (privately published). He lives in Massachusetts.*

He lived in a small apartment in an assisted living facility, with an Escher illustration on his door, and had converted most of his single room into a working space. I recall two large bookcases, one serving as a partition between his sleeping area and a slightly larger office area, and the other housing his large personal collection of the books he authored. At the time he was working on the revision to one of his books on scientific skepticism, and two of his many file drawers had been retrieved from his offsite storage area and took up most of his working surface area. Martin did have a computer; not for email or Internet access but for referencing his many *Scientific American* articles and books, the index of which was on CD-ROM. He found it humorous that the other residents at the center, primarily women, were very curious about what kind of work he did, with visitors always coming and going; he had a private table in the dining room and I think he took pleasure in being that quiet, mysterious guy down the hall.

For those who sent Martin inscribed copies of their books as gifts, you may not want to know that unless he was using them on a project, they invariably got recycled to the next visitor showing interest in the subject. I walked away with a few of the books that well-wishers had given him, and I'm sure the copies I left of my *Square Alice* and a Lewis Carroll comic book bibliography I helped to edit later found their way into someone else's collection. We talked for most of the day and had dinner together and took photos before my wife and I left for Iowa.

Uncharacteristically, Martin agreed to having a meeting of the LCSNA sometime in the near future in Norman, to celebrate his significant contributions to Carrollian scholarship and popularity. Unfortunately, he wrote to me a few months later to ask that we change our plans, due to concerns about his health.

I can't tell you how many subjects we touched upon that day, from games and puzzles to his wife's former collection of door stops (including a few rare Alice ones)—he even found the business card of the auction house handling its sale. I do recall his smile when I mentioned the happy coincidence that our favorite Carrollian number (42) and our common interest in palindromes meshed in its binary equivalent: 0101010. He said he would include it in the next set of annotations but, alas, it will remain only as an annotation to my fond memories of a man who was intellectually active until the very end.

My Correspondence with Martin Gardner

EDWARD WAKELING

The first letter from Martin Gardner in my correspondence file is just an envelope postmarked November 7, 1988. I received it as treasurer of the Lewis Carroll Society (U.K.), and it probably contained a check for Mr. Gardner's subscription. Normally I would throw envelopes away, but since it was personally addressed to me and it had Martin Gardner's address printed in the top left corner, I kept it.

About a year later I received a proper letter, dated November 13, 1989. It was short and to the point, and immensely helpful:

> Dear Mr. Wakeling,
> Dover let me see your MSS on L.C. puzzles and asked my opinion. I strongly urged them to make you an offer for it. You did a splendid job.
>
> Best,
> Martin Gardner

Edward Wakeling did postgraduate studies at Christ Church, Oxford, in the early 1980s, originally training as a teacher of mathematics. He is a former Chairman of the Lewis Carroll Society (UK), and a Carroll collector, scholar, researcher, and author, whose published works include Lewis Carroll, Photographer *with Roger Taylor and* Lewis Carroll & His Illustrators *with Morton Cohen. He recently edited the first unabridged edition of* Lewis Carroll's Diaries, *published in ten volumes.*

That manuscript, which eventually saw print as *Lewis Carroll's Games and Puzzles*, was a result of my having worked with schools, and encouraging them to adopt some recreational mathematics as part of their curriculum. With a group of teachers we put together *Alice's Adventures in Curriculum-Land* and my employer, Bedfordshire County Council, printed and distributed copies to all schools in the county. Soon after this I was invited to be a trustee of the newly formed Lewis Carroll Birthplace Trust, and I wanted to make a tangible contribution to their fundraising. I rewrote *Curriculum-Land* with a lot of additional information, concentrating on the mathematical and linguistic puzzles invented by Lewis Carroll, and submitted the manuscript to Dover Publications in New York.

I was aware of their range of books on recreational mathematics, some within my own library. Their books seemed to stay in print for decades. I was not concerned about payment or royalties as long as they were prepared to make a donation to the Birthplace Trust. Martin Gardner's support was most welcome. I have since seen his report to Dover Publications; a copy was sent to me by Martin, in which he wrote, "I think Dover would do well with this because there is sure to be rising interest in Carroll in 1990. It should see the appearance of my *More Annotated Alice*, which Random House intends to promote." He goes on to say, "Because Tenniel's illustrations are public domain, you might consider adding some of them to the book at spots where Wakeling quotes from an *Alice* book." Dover accepted the book, and Alexander Macmillan, the Earl of Stockton, chairman of the Trust, whose father had been prime minister and whose great-great-grandfather published the original *Alice's Adventures in Wonderland*, wrote a foreword. The book was published in 1992. A generous sum went to the Trust, which was later able to purchase the land on which Daresbury Parsonage once stood, thus protecting the site for the future.

My gratitude to Martin knew no bounds. A friendship developed that resulted in a steady correspondence between us, but, sadly, we never met in person. From time to time, Martin would send me a query that he was struggling with, seeking my help. For example, in June 1990 he wrote about a question that had arisen while preparing his notes for *More Annotated Alice*: "On page 60 of *The Lewis Carroll Handbook* (current edition), are listed the changes that LC made on his prefatory poem for the second *Alice* book. In the second line of the fifth stanza, he changed 'That lash themselves' to

'The storm wind's moody.' But there is a word missing here. The corrected line reads 'The storm wind's moody madness.' But the original line, in LC's handwriting, surely is not 'That lash themselves madness.' It was probably 'That lash themselves *to* madness' or 'That lash themselves *in* madness.' I have been unable to determine where Carroll's handwritten copy of the poem is located. Would you or one of your Carrollian friends know how the line originally read?" I replied on 2 August 1990, apologizing for my delay:

> I contacted Christ Church to see whether the manuscript version of the poem is among their holdings. It is not. I have been through all the manuscript items in both the Library and Archives at Christ Church, and I am 99% certain that the poem is not there. The Assistant Librarian says that it is not listed in their catalogue of Carrollian items. Re-reading the Handbook, it seems to suggest that the original version is in the Amory Collection at Harvard; have you tried there?

My letter went on to tell Martin about the illustration plan for *Through the Looking-Glass* I found at Christ Church. I explained that the plan gave details of the change in frontispiece and the demise of the "Wasp" chapter. I said: "Captions for the proposed illustrations are given together with tentative page numbers. The change from black to violet ink helps to date some of the changes. I intend to work on my paper [subsequently printed in *Jabberwocky* Vol. 21, No. 2] very soon, and will be happy to send you a copy if you think it might be relevant to your work." I ended my letter by saying: "I send you my very best wishes, especially for a successful publication of *More Annotated Alice*. The original book has been a constant source of information for me; it was one of the first books I bought for my collection."

A week later I received his reply:

> I certainly didn't expect you to spend so much time looking for Lewis Carroll's corrections to that poem, but I do appreciate it. No, I haven't tried the Harvard collection, but now I have checked galleys on my *More Annotated Alice*, so it is too late to make any changes or additions. Random House has ordered a first printing of 30,000 (this November) which augurs well for sales.

In the same letter he spoke about his "crazy novel," *The Flight of Peter Fromm*, in which a "minister cracks up during an Easter sermon and starts reciting 'Jabberwocky'!"

On 21 August 1990, I sent Martin an advance copy of my article on the illustration plan for *Through the Looking-Glass* even though it was too late to use anything in *More Annotated Alice*. He sent me his thanks on August 28. His letter also contained details of Justin Schiller's new book, *Alice's Adventures in Wonderland: An 1865 Printing Re-Described*. The letter ended with a discussion of the King's "whiskers" and the image of the White Knight in *Through the Looking-Glass*, and Tenniel's large handlebar moustache.

My next letter is dated December 11, 1990, and the text is as follows:

Dear Martin,
I hope it is not too forward to address you thus. You know how reserved we English can be; anyway, you signed your last letter to me in this manner, so I thought I would be a little more friendly in my greetings this time, and wait for your response.

> To my shame, I note that your last letter to me is dated 28 August. I apologise for the delay in replying. Since that time, your *More Annotated Alice* has come out, and the publishers have sent me (probably at your request, for which I am grateful) an un-bound copy to whet my appetite. It is a beautiful book, and I will enjoy revisiting your work and adding to my own knowledge the fruits of your more recent research in matters Carrollian.
>
> Thank you for your advice about Dover books. In the meantime, they have settled the payment to the Lewis Carroll Birthplace Trust as arranged with me, and now I await the next stage of the publishing process. I am told that they intend to publish the book in the Spring.
>
> Thank you for sending me Justin Schiller's address. I had already ordered his book, and it has now arrived. Again, a marvellous contribution to Carrollian research, and very finely done.
>
> Your comments about the King's whiskers has surprised me; I thought that this referred to a beard rather than sideburns, but you are right. The *Oxford Dictionary* describes whiskers as hair growing on the cheek, not on the chin! Do you know when Tenniel grew his handlebar moustache? All the pictures I have seen of the man show his luxurious growth. You suggest that this was after 1872. You will know that M. H. Spielmann in *The History of "Punch"* (p. 329)[1] suggests that Tenniel took his colleague Horace "Ponny" Mayhew as his model for the White Knight, although Tenniel is quoted in the book as saying "the resemblance was purely accidental, a mere unintentional caricature." There is more in this book about moustaches worn by members of the *Punch* staff (pp. 423–4) which seems to suggest that Tenniel had his handlebars as early as 1865!

I ended by thanking him for his membership subscription, which I passed to the new treasurer, and sent him best wishes for Christmas and the New Year.

While attending the Second International Lewis Carroll Conference at Winston-Salem, North Carolina, in 1994, I was told by Stan Marx and other representatives of the LCSNA that Martin Gardner's contribution to the pamphlets project on puzzles and games had been rejected because it was not in the required format. I was asked if I would do the book myself. Of course, I declined; such an action would have compromised my friendship with Martin. In the end, Martin published his work as *The Universe in a Handkerchief* in 1996, and I sent him a complimentary postcard soon after the book came out.

Later I reviewed the book for the *Lewis Carroll Review* and wrote to Martin on November 17, saying: "I enjoyed the book very much, and I'm grateful for the mentions of my work. I get the impression that you do not have a copy of my second book, so I have pleasure in enclosing one for you. As I say in the review, I have been collecting more examples of Carroll's puzzles and games, and have sufficient material for a third volume. The only problem is time to write the book. As you know, I am tied up with the diary project for the next few years, but I have been able to work on other projects in odd moments, and perhaps the third puzzle-book will see the light of day before too long." The third puzzle-book did not come to fruition. My letter went on to describe my research scholarship at Princeton University, and I sent Martin some mathematical examples of *Memoria Technica* that I had found among the Parrish Collection.

Martin wrote back to me on November 27, 1996, thanking me for the review. He went on to say: "I envy all your research, and eagerly await reading some of your discoveries, especially those unpublished math manuscripts. I agree that LC's math ability is underrated, and continue to be amazed at how much of his work you are finding." The letter went on to mention the nonsense published by Richard Wallace as *Jack the Ripper: "Light-hearted Friend"* and Martin included a review he had copied from the November issue of *Harper's Magazine*. He added: "Surely it is a hoax!"

I did not reply until May 29, 1997—my time for personal correspondence was limited at that time as a result of pressures in my working life. I commented on the Wallace book by writing: "Thank you for the photocopy of the review in *Harper's Magazine* of Richard Wallace's outrageous book, *Jack the Ripper, 'Light-hearted Friend.'* I corresponded with this man many years ago when he wrote an equally insane book called *The Agony of Lewis Carroll*. Do you know it? He used the same technique of anagrams from Carroll's works which he adopts in this current book. I tried to point out to him that if you chose 50 or more letters, then you can make any message you like from them. The number of combinations is enormous. But he wouldn't have it." I spoke about my review of the book commissioned by the Cloak and Dagger Club and published in their journal, *The Ripperologist*, in which I totally discredited the book and outlined a catalogue of errors and flaws in Wallace's evidence. My letter went on to discuss other Carrollian matters including "white stone days," Dodgson's view of Darwinism, photographs, presentation copies, correspondence numbers, and Dodgson's library. I ended my letter by asking whether Martin was working on any further Carroll books with 1998 rapidly approaching.

The last letter I received from Martin is dated June 5, 1997. I reproduce it in full below.

> Dear Edward,
>
> Congratulations on Diary 4. I eagerly await it, and of course also Vol. 5. If you will send me a copy of your review of the Ripper book I will be most grateful. I found it incredible that a magazine like *Harper's* would give it publicity.
>
> I own no presentation copies of any LC books. I must confess that I'm not a LC collector. The books I have by and about him include nothing rare. I did once own a copy of his book on determinants, signed "from the author," but I donated it to a library in Canada.
>
> I didn't know about the book of LC letters you are doing with Morton. It should be a major contribution to the growing LC literature. Three new biographies?! It's hard to believe there is that much new to say! I have no further LC book in mind. I would like to update my *Annotated Snark*, but Penguin reissued their edition last year without giving me a chance to add new notes.
>
> I enclose my source for using chalk to mark calendars as white stone days, and a cartoon someone just sent me.
>
> Penguin has just published a paperback edition of my latest book, *The Night is Large*. I mention it because it reprints my introduction to *The Annotated Alice*.
>
> All best,
>
> Martin

1 Spielmann, M. H. *The History of "Punch"* (London: Cassell and Company, Ltd., 1895)

Editing Martin

ROBERT WEIL

My first recollections of Martin come from the 1980s when I was a young magazine editor and he a renowned essayist, columnist, and prominent world intellectual, the last being a term he might, characteristically, find too grand. At that time, Martin would regularly contribute pieces and puzzle ideas to *Omni* magazine, where I was Features Editor and ran a small book division. Needless to say, I was in awe of him, given his reputation, and I recall that the magazine's Games Editor, Scot Morris, treasured his advice and would use him regularly as a source for ideas. While I had nothing to do with the "Games" column, I did work with Martin on a few articles and would interview him on occasion—when I wanted to be grounded on what was indeed fact as opposed to pie-in-the-sky, futuristic fantasy. Even though I was only in my twenties, I was stunned at how patient and kind Martin was, and this was often quite a contrast with many of the world-class scientists and writers I had to deal with.

Robert Weil, who began his storied editorial career in 1978, has been an executive editor at W. W. Norton since 1998. In addition to the ten books he produced with Martin Gardner, he has worked with some of the most successful and revered authors in the world, such as Edward O. Wilson, Nadine Gordimer, J. G. Ballard, and Paul McCartney. His books have won multiple awards, including the National Book Award, and the Pulitzer Prize for Annette Gordon-Reed's The Hemingses of Monticello.

It was not until I was already at St. Martin's Press in the mid-1990s that I would commission my first book with Martin, a superb essay collection called *The Night Is Large*, which was published in 1996, and which remains, perhaps, Martin's most important essay volume. The book received extraordinary comments and remarkable reviews, and my favorite one was from Dennis Flanagan, the former editor of *Scientific American*, who wrote that "If there is a God (a question taken up in *The Night Is Large*), He owes Gardner a medal."

Two other recollections stand out from that first collection. First, I recall getting a call from a very brusque Stephen Jay Gould, then at Harvard, who was at the top of his powers, as well as a huge bestselling author. He chastised me at first, telling me that he never gave blurbs, but then said that the only exception he would make would be for Martin, about whom he remarked, "For more than half a century, Martin Gardner has been the single brightest beacon defending rationality and good science against the mysticism and anti-intellectualism that surround us. He is also one of the most brilliant men and gracious writers that I have ever known." Gould actually dictated the quote to me on the phone, making me repeat the commas, but because the comment was so heartfelt and because it was the only time I actually spoke with Gould, I was ecstatic about the conversation.

My other recollection of that first collaboration was in working with Martin and Charlotte on Christmas Day, either in 1994 or 1995. We were going through nearly a hundred essays, winnowing them down, making selections, perhaps writing copy and adding introductions, and I vividly recall that it was the only time I spoke with Charlotte at length. I think Martin and Charlotte, nonbelievers themselves, were the only ones who did not find it strange that I was laboring that yuletide day from morning to night, but it was clear that they were doing the same. I only wished I had known Charlotte better, since she was engaging and fiercely intelligent in the kindest of ways, and she seemed to very much know and love the man she had married many years before. Over the years, Martin and I would discuss Charlotte on many occasions, and he suffered, I recall, a terrible depression after her death, and I would make a point to call him very often just to make sure that he would not, as I sometimes feared, go over the abyss. I only knew that he had recovered several years later when he began writing books for me again.

Over the years I ended up editing ten books with Martin, including the aforementioned *The Night Is Large*, as well as the revised edition of *The Whys of a Philosophical Scrivener*, the delightful *Visitors from Oz* novel, which reflected Martin's love and obsession with the L. Frank Baum stories, as well as a complete rewrite of Sylvanus Thompson's *Calculus Made Easy,* in which I had suggested to Martin that he take this classic calculus work, which had been written during the era of World War I, and update it in contemporary terms. I was amazed that Martin could take a difficult mathematical work that was over seventy years old, and completely redo it in modern parlance. The book did especially well, and this was the fourth of four books that I handled with him at, appropriately, St. Martin's Press.

The first new book that I signed up when I went to W. W. Norton & Company, in December of 1998, was *The Annotated Alice: The Definitive Edition*, in which Martin combined the annotations from his 1960 classic with those from his 1990 *More Annotated Alice*, and then he added yet a third complement of notes, forming the one work for which he is perhaps best known. It was a fortuitous beginning for me at Norton, for the book has now sold well over 100,000 copies, is one of our strongest backlist titles in hardcover, and is simply the book that keeps giving. It is in literary terms a classic work, for there is no one who was more brilliant and more accomplished a critic of Lewis Carroll than Martin himself, who seemed to channel Carroll's thoughts and recondite ideas for the twentieth and twenty-first centuries.

I'm sure I could describe each book I did with Martin, so it's best that I merely mention the others' titles, including *Did Adam and Eve Have Navels?*; *The Colossal Book of Mathematics*; *Are Universes Thicker than Blackberries?: Discourses on Gödel, Magic Hexagrams, Little Red Riding Hood, and Other Mathematical and Pseudoscience Topics*; *The Colossal Book of Short Puzzles and Problems*; and finally, a completely revised and new edition of *The Annotated Snark*, which would, sadly, be our last collaboration. He was so excited about this last book, given his affection for the Bellman and the Snark and Carroll's most strange creation, and I would work with him on the phone, cutting out annotations and additions from various books and articles, and conferring with him all the while (Martin felt that we should include an advisory suggesting that the reader purchase red cellophane tape to observe some hidden detail in an accompanying drawing, but this was difficult if not impossible, for it proved challenging to incorporate his ideas for a new, twenty-first-century audience, which did almost everything online and would have been hard-pressed to buy acetates or red cellophane in a store).

It's worth noting that a man as mathematically gifted and technologically brilliant as Martin refused to countenance cyber-technology—to send email or use a personal computer. In many ways, he remained—almost in the tradition of Emerson or Thoreau, a great nineteenth-century scholar and writer who lived solely through the meanings and ambiguities of the written word. To his profound understanding of language he also brought an unparalleled mathematical sense, and this singular combination was perhaps a one-of-its-kind sensibility in the wide canyons of American culture. A few others, of course, possessed Martin's genius, but none could translate that knowledge in a way that the popular reader could understand, and it was this perceptive gift—that of marrying these two worlds—that made his contributions, in my experience, completely unprecedented.

It is clear for anyone who really knew Martin that he was quite a polymath. His brilliance was homespun, and he was as comfortable with folktale, badinage ("Casey at the Bat"), and wordplay as he was with high literature and theoretical or applied mathematics. I often thought he had the unique talents and superb writing ability to have been a lifelong editor at *The New Yorker*, for he often reminded me with his catholic tastes and literary erudition of John Updike, but Martin remained an Oklahoma Sooner to the last, pursuing—with Charlotte as his finest muse—his intellectual life just in the way he had created it, celebrating the most common strands of American popular culture as well as the literary ideas emanating from the academy, especially from the University of Chicago, where Robert Maynard Hutchinson's philosophy had profoundly influenced him as a young man. There was a remarkable and sovereign life force that distinguished Martin throughout his seventy-plus-year career, but there was also a modesty—often reflected in his chuckles and long pauses—which made him so different from the other geniuses of his ilk.

I think it's appropriate to close by stating that, given his output and influence so far, none of us have heard the last, to quote Douglas Hofstadter, of "one of the great intellects produced in this country in this [last] century."

FROM THE *Knight Letter*

Well, You Know

MARTIN GARDNER

James B. Hobbs, professor *emeritus* of business administration and associate dean *emeritus* of the College of Arts and Sciences at Lehigh University in Bethlehem, Pennsylvania, recently called my attention to an aspect of both *Alice* books that I had not noticed before. He was struck by the unusual frequency with which Alice and 23 other of her companions needlessly interject "you know" into their conversations.

In recent years, the use of "you know" in the United States has become a compulsion among many young people, athletes, talk-show hosts and their guests. In some cases, it seems impossible for them to utter a sentence without interjecting "you know." Although unlikely that American children picked up the habit merely from reading the *Alice* duad, the question arises: Was this a common practice of dialogue in Carroll's day? Or might another explanation exist?

First the facts; then two conjectures. Alice needlessly interjects "you know" 31 times in the two volumes. (The twenty times the phrase was used as a question or as a declarative statement are excluded.) The phrase is used a total of 86 times in the two books: five times each by the White Knight, Humpty Dumpty, and the Red Queen, and four times by the White Queen. Nineteen other characters use the phase between one and three times. After being confronted with so many "you knows," it was most gratifying to observe the Caterpillar's retort in *Wonderland* to Alice bemoaning her changing size so often, "you know": "I *don't* know!"

Hobbs also tabulated the frequency that "you see" is used by Alice and her companions in the two volumes. Although Alice uses this phrase only twice—once in conversation, again with the Caterpillar, where he says: "I don't see!"—the White Knight uses it eight times, the narrator seven, and seven other characters once or twice. Total "you see" interjections is 26. A third phrase, "of course," is used 35 times: six by the Mock Turtle, four by Humpty Dumpty, and between once and three times by fifteen other characters (including Alice thrice).

Two conjectures arise as to why Carroll interjected these three phrases so frequently. *One*: such phrases might have been the norm in talking with or between small children, particularly little girls, during the middle and late nineteenth century. *Two* (and Hobbs suggests this is the more plausible): Carroll (or C. L. Dodgson) was a teacher/professor and a logician/mathematician. Instructors frequently use "you know" and "you see" to clarify (hopefully or in fact) the point under discussion. And it is also not uncommon to emphasize a point (whether complex or "simple") with an "of course" thrown in. To boot, bred into the logician/mathematician is the discipline to develop detailed proofs for theorems and constructs, which are often terminated with the super-flourish: "*obviously* such and so is the conclusion or insight!" Nary an "obvious" appears in either tome, suggesting that Carroll may have replaced a natural (or trained) tendency to use the word with the less harsh and intimidating "you know," "you see," or "of course."

I would greatly appreciate learning if the frequency of these three phrases has been noticed by other Carroll scholars, and if so, their thoughts on the matter.

[*Hobbs's chart tabulating the characters and the occurrences of these phrases was attached. An email forwarded to Edmund Weiner, the Principal Philologist of the* Oxford English Dictionary, *inquiring about the correct rhetorical term for unwarranted interjections such as "you know" received this reply: "I think the traditional word is* expletive, *defined by* OED *as 'serving merely to fill out a sentence, help out a metrical line, etc.' I think that in modern grammar a more accurate term is* filler, *defined by the* Oxford Dictionary of English Grammar *as 'A word, usually outside the syntax of an adjoining clause, that serves to fill what might otherwise be an unwanted pause in conversation.' Another term for this is* pragmatic particle."*– Ed.*]

The Final Annotations

MARTIN GARDNER

Gardner's two last sets of annotations, here integrated together, were first published in Knight Letter *2, issue 5, no. 75 (Summer 2005) and issue 6, no. 76 (Spring 2006). – Ed.*

INTRODUCTION

Since Norton published in 1999 what they called the "definitive" *Annotated Alice*, numerous readers have written to propose new notes, and other good suggestions have been made in books and periodicals. In brief, the new edition is far from definitive, a goal it surely will never reach. Rather than add new notes to another revised edition—that would be unfair to purchasers of the present edition—I decided to send to the *Knight Letter* a supplement containing more notes and a few trivial corrections.

The first edition of *The Annotated Alice* was published by Clarkson Potter in 1960. It was followed thirty years later by *More Annotated Alice*, with art by Peter Newell, published by Random House. The present Norton edition combines the text of both books, with many fresh notes tossed in. The past few years have seen a continuing flood of new books and articles about Lewis Carroll and Alice. The number of Carroll biographies now exceeds twenty, the best (in my opinion) by Morton Cohen.[1] The Lewis Carroll Society in England publishes three periodicals, *The Carrollian*, *Lewis Carroll Review*, and *Bandersnatch*. The Lewis Carroll Society of North America publishes the *Knight Letter*. Other publications come from similar groups in Canada, Australia, and Japan.

Many pages would be required just to list new illustrated editions of the *Alice* books, as more books by or about Carroll appear every year. I have written *The Universe in a Handkerchief*, about his original puzzles and games; annotated *The Hunting of the Snark*, and *Phantasmagoria*; and penned introductions to *The Nursery Alice*, *Alice's Adventures Underground*, and the first volume of *Sylvie and Bruno*.[2]

Books and papers by Morton Cohen continue to reveal surprising new information. A raft of plays, musicals, films, even ballets keep turning up on stage and screen. New translations of *Alice* are being made throughout the world, especially in Russia and Japan. (For the proliferating Russian literature, see Maria Isakova's fine article in *Knight Letter* 74, Winter 2004.)

So much for the bright side of the ongoing Carroll Renaissance. There is a darker side. I refer to the burst of criticism by a small group of scholars known to outsiders as "revisionists," and to themselves as "Contrariwise: The Association of New Lewis Carroll Studies." (The term comes, of course, from the Tweedle twins.) They even have a Web site called *Looking For Lewis Carroll*.[3]

The purpose of this "new wave" of criticism is to explode what its leaders call the "myth" of Dodgson as a devout Anglican who had almost no interest in boys or mature women, instead concentrating his affections on attractive preadolescent girls, with a special love for young Alice Liddell. According to Professor Cohen's theory, Dodgson actually hoped that someday he might marry an adult Alice.

"Contrariwise!" shouts Karoline Leach, who started the revisionist movement. In her explosive book *In the Shadow of the Dreamchild: A New Understanding of Lewis Carroll*, she does her best to demolish his saintly image. To replace it, she depicts him as a normal heterosexual who used his child-friendships "as the cleanser of his grubby soul."[4] It is impossible to believe Leach's contention that not only did Dodgson engage in adultery with Mrs. Liddell, Alice's straightlaced mother, but that he had similar affairs with other adult women.[5]

Leach's claims strike Professor Cohen and many other Carrollians, including me, as on a level with the absurd premise in Dan Brown's *The*

DaVinci Code[6] that Jesus was married to Mary Magdalene, who appears in DaVinci's "Last Supper" as sitting to the right of the Lord. Leach's revelations are almost as preposterous as a book, actually published years ago, proving that Carroll was Jack the Ripper, or another idiotic work exposing Queen Victoria as the true author of the *Alice* books!

For some comments on the "Contrariwisers," see Morton Cohen's slashing article "When Love Was Young: Failed apologists for the sexuality of Lewis Carroll" in the *Times Literary Supplement*.[7] Cohen bases his attack on Leach's book and on her two articles in earlier issues of the same publication.[8]

I give Leach Brownie points for calling attention to a peculiar long-forgotten short novel, *From an Island* (1877, reprinted 1996). The author was William Thackeray's daughter Anne. In an article in *The Carrollian*,[9] Leach asserts that Anne's novella is a roman à clef, its main characters based on then living persons such as Tennyson and the artist G. F. Watts. The book's central figure has the strange name of George Hexham, a young photographer possibly from Christ Church College, Cambridge.[10] During a visit to the Isle of Wight he falls in love with the novella's heroine, Hester, and she with him. Leach maintains that Hester is a thinly disguised Anne, and that Hexham is—tighten your seat belt!—none other than our Mr. Dodgson.

Thackeray does a cruel hatchet job on Hexham. He is portrayed as tall and handsome (no trace of a stammer), but selfish, pushy, self-centered, obnoxious, easily angered, and rude to everybody including Hester. His hair is "closely cropped," unlike Carroll's long, flowing locks. He treats Hester with callous indifference while he flirts shamelessly with another woman. At the story's end the two lovers have an improbable reconciliation.

In the article, titled "Lewis Carroll as Romantic Hero," Leach discloses that Dodgson owned a copy of *From an Island*, and in a letter praised Anne's writing style as unusually "lovely." On October 5, 1869, he briefly mentions in another letter that he "met" Anne at a dinner party. Leach assumes, with no evidence, that "met" does not here mean he met Anne then for the first time. She conjectures that he met her years earlier, but that a diary entry on such a meeting must be in one of the diary's lost pages. It seems to me that Dodgson, who thought so highly of Anne's work, would somewhere have dropped a hint that he and Anne were more than just casual friends. It is also strange that she identifies Hexham as a photographer, not as the author of a famous children's book. (The first *Alice* was published three years earlier than Anne's novel.)

Keith Wright, discussing *From an Island* in a letter to the editor,[11] argues persuasively that Hexham, unlike other characters in the book, was an entirely fictional creation. Dodgson, he writes, did indeed visit the Isle of Wight on three occasions. Mrs. Tennyson kept a journal that mentions the last two visits, but makes no reference to Anne and Dodgson being on the island at the same time. Dodgson's first visit is recorded in his diary for 1864. There is no mention of Anne. I find it a huge stretch to suppose that a youthful Dodgson would have the ugly personality of Hexham without a record of someone else's similar impression. Leach suspects that a romantic episode with Anne underlies his later poems about the love of a woman.

Leach missed a subtle Carrollian clue, which I recently discovered. It is based on what wordplay enthusiasts call "alphabetical shifts." Move each letter of GH, the initials of George Hexham, back four steps in the alphabet. You get CD for Charles Dodgson! And if you move CD forward another four steps you arrive at KL, the initials of Karoline Leach. Another bit of numerology links LC to GH. For each letter in LC substitute the number of its position in the alphabet using the code A = 1, B = 2 and so on. The letters of LC have the values 12 and 3, which add to 15. The same sum is obtained when you do the same with GH.

(Of course that last paragraph was entirely tongue in cheek!)

At any rate, Leach deserves credit for uncovering a bizarre and still unresolved literary mystery.*

Matthew Demakos, in a letter to *The Carrollian,*[12] calls attention to six scholars who speculated on the identity of characters in *From an Island*. It appears from their conflicting opinions that the novella may not be a roman à clef after all! There is no agreement about the then living counterparts of Anne's narrative. For example, Tennyson could be portrayed as either St.

* Hexham is a town in Northumberland, a northern county of England. Can any reader provide a good explanation of why Anne would apply this name to George Hexham? Is it possible that there was a photographer at Cambridge University who came from Hexham? [*Is it of interest that* HEX- *(Gr.* ἕξ, *six) is half of* DOD- *(Gr.* δώδεκα, *twelve)? – Matt Demakos*]

Julian or Lord Ulleskelf. St. Julian could be based on Browning or Watts as well as Tennyson, and so on for other characters.

The novella, Demakos reveals, first appeared in three installments in *Cornhill Magazine* (1868–69), before any record of Carroll having met Anne. In the novel Hexham sends a letter from Lyndhurst, with no explanation of why he was there. All very mysterioso.

NEW NOTES AND CORRECTIONS

The page numbers are for "The Definitive Edition" of *The Annotated Alice* (Norton, 2000). Pagination is different in the Penguin British edition.

xvii. Place a [1] for an end-note after "white stone" in line 13 from bottom.

xxii. Add an end-note:

1. For a history of the ancient practice of marking a special day or event with a white stone, see Kate Lyon's essay "The White Stone" in *Knight Letter* 68 (Spring 2002).

11. Place a [1a] for another annotation at end of second paragraph.

12. Add an annotation:

1a. Professor D. T. Donovan, University College London, reminded me that the White Rabbit's pink eyes identify him as an albino.

Add to the first paragraph below the quotation in the footnote:

A subtle indication of Carroll's influence on L. Frank Baum is the fact that the first word of the first *Oz* book is "Dorothy." Linda Sunshine has published beautifully illustrated tributes to both authors: *All About Alice*[13] and *All About Oz.*[14] She, Angelica Carpenter, myself, and many others are among those who are both Carrollians and Ozians.

14. Place a [5a] for an annotation after the words "you know" in second paragraph, line 7.

Add an annotation:

5a. This is the first time Alice says "you know" as a needless interjection. James B. Hobbs surprised me by pointing out that Alice says "you know" more than thirty times in the two *Alice* books. Other characters say "you know" more than fifty times! These numbers do not include "you know" when used normally, just when used as a meaningless phrase.

Both Alice and the characters she meets repeat "you know"s like many of today's American youths even after they become adults. Is it possible that "you know" was a similar speech fad in Carroll's day? Hobbs found it gratifying that when Alice says "you see" (another of her favorite expressions) to the Caterpillar he replies "I don't see," and when she later says "you know," the Caterpillar remarks, "I *don't* know." See my article "Well, You Know..." in *Knight Letter* 65 (Winter 2000). [*Reprinted in this present volume. – Ed.*]

Add an annotation:
5b. Brian Sibley noticed that in Tenniel's picture (p. 22) of the White Rabbit trotting down the hall, no lamps are hanging from the ceiling.

15. Add a [5b] at the end of the top paragraph.

23. Add to note 4:
See also "Alice in Mathematics," by Kenneth D. Salins, in *The Carrollian* (Spring 2000).

28. Add to note 10:
See Brian Sibley's delightful essay "Mr. Dodgson and the Dodo," in *Jabberwocky* (Spring 1974). He quotes Will Cuppy, "The Dodo never had a chance. He seems to have been invented for the sole purpose of becoming extinct and that was all he was good for."

30. Line 4 of note. Change the year to 1848.

32. Add to end of note 2:
At the end of the previous chapter's Note 10, I mentioned the surprising appearance of an ape in Tenniel's pictures of the creatures present at the Caucus Race. Carroll himself had introduced the ape in the sketch he made for *Alice's Adventures under Ground.* Because the ape is nowhere mentioned in the text of that book or in the first *Alice* book, critics have understandably wondered why Carroll added an ape and allowed Tenniel to do likewise. The consensus is that the ape's presence reflected public controversy over Darwin's theory of evolution.

Did Carroll believe in evolution? It has been said that he did not. I'm not so sure. In his diary (November 1, 1874) he expresses his admiration for a book by the British zoologist St. George Jackson Mivart:

> Not being well, I stayed in all day, and during the day read the whole of Mivart's Genesis of Species, a most interesting

> and satisfactory book, showing, as it does, the insufficiency of "Natural Selection" alone to account for the universe, and its perfect compatibility with the creative and guiding power of God. The theory of "Correspondence to Environment" is also brought into harmony with the Christian's belief.

Now Mivart, a student of Thomas Huxley, fully accepted an ancient earth and the evolution of all life from single-celled life forms. However, like today's proponents of "intelligent design," he argued in his book that God created and guided the evolutionary process and at some moment of history infused immortal souls into ape-like beasts.

In 1900, the Catholic Church excommunicated Mivart for heresy. In recent years, the Vatican has officially endorsed Mivart's intelligent design view. For the sad story, see Chapter 9 of my *On the Wild Side.*[15]

36. First line of note. Change "New York" to "New Jersey." (This was corrected after the first edition.)

    Insert the following between the two paragraphs of Note 4:
    See "A Tail in a Tail-Rhyme," by Jeffrey Maiden, Gary Graham, and Nancy Fox, in *Jabberwocky* (Summer/Autumn 1989), and its references.

39. Add to note 4:
    Correction: Gordon Claridge wrote from Oxford University to say that this phrase is heard only in Scotland.

48. Add to note 2:
    See an earlier article, "Alice, Who Are You?" by Fred Madden, in *Jabberwocky* (Summer/Autumn 1988). He also mentions that Carroll owned a copy of MacKay's book, with its chapter on "Popular Follies in Great Cities."

64. Add to note 4:
    John Shaw, writing on "Who Wrote 'Speak Gently'?" in *Jabberwocky* (Summer 1972), also gives a history of the controversy in which he played such a major role. He provides a bibliography of 56 publications of poems that begin "Speak gently." Carroll's parody, he concludes, "may well be an echo of all of them rather than any one of them."

Carroll's parody has been set to music by Alfred Scott Gatty. The score, undated, is reproduced in *Jabberwocky* (Winter 1970).

65. Add to note 5:
Tenniel's picture of Alice holding the pig is one of his very few illustrations showing Alice full face, looking straight ahead. Note the foxgloves on the left.

68. Add to note 10:
Ferdinand J. Soto, in *The Carrollian* (Spring 1998) suggests that Alice left a straight road and for a short distance followed a circular path that put her back on the straight road. Of course the simplest explanation is that Tenniel failed to notice that Alice had "walked on."

70. Add to note 1:
"Carroll never actually describes any of his characters," writes Linda Sunshine in the *Knight Letter* 74 (Winter 2004). "So illustrators are really free to use their wildest imaginations to create their very own Wonderland."

102. Add to the quotation from Mary Howitt's poem at the top of Note 3:
See "The Contribution of Mary Howitt's 'The Spider and the Fly' to *Alice's Adventures in Wonderland*" by Chloe Nichols, in *The Carrollian* (Spring 2004). Howitt's entire poem of seven stanzas is reprinted in the issue's Appendix.

103. Add to the top note:
For a good account of the traditional tunes for many of the songs in the two *Alice* books, as well as melodies for songs written by later composers, see Armelle Futterman's article "'Yes,' Said Alice, 'We Learned French and Music.'" in *Knight Letter* 73 (Spring 2004).

107. Add to note 8:
David Lockwood, writing on "Pictorial Puzzles in Alice" (*The Carrollian*, Autumn 2004), makes a good case for the appearance of all five opening ballet positions in illustrations for the first *Alice* book (Tenniel's father was a dancing master). The lobster is in the first position (p. 106); the knave of Hearts is in second position (page 88); the fish (page 58) is in third position; and the frog, in the same drawing, is in the fifth. Alice, in the picture on page 32, is in fourth position.

That these are not coincidences, Lockwood argues, is supported by the near absence of any ballet positions in Tenniel's art for the second *Alice* book. Only the first position turns up in the stances of the Tweedle brothers. Lockwood ends his article with some interesting speculations about the origin of the command "Off with her (or his) head!"

120. Add to the note:

Adams himself has denied that he had Carroll in mind when his computer Deep Thought answered the "ultimate question." It was no more than a joke—a random number that popped into his head. (See *Knight Letter* 75, Summer 2005.)

For more speculations about 42, see the three articles on the topic in *Jabberwocky* (Spring 1993) and musings by Charles Ralphs and others in *Jabberwocky* (Winter/Spring 1989), Ellis Hillman's "Why 42?" in *Jabberwocky* (Spring, 1993), and *The Alice Companion*, by Jo Elwyn Jones and J. Francis Gladstone (New York University Press, 1998, 93–94).

Yuriko Kobata wrote to tell me about the following correlation she had discovered. For each letter in DODGSON substitute the number of its position in the alphabet (A = 1, B = 2, etc.). The sum of the eleven digits is 42.

123. Add to note 5:

It was suggested that in the picture on the left, also in Tenniel's frontispiece, the Jack is not the Jack of Hearts but the Jack of Clubs. Why? Because tiny emblems on Jack's tunic look like clubs. However, if you check the Jack of Hearts in any modern deck you'll find the same emblems. They are not clubs but three-leaf clovers—the Irish shamrock, widely taken by Irish Christians to be a symbol of the Holy Trinity.

124. Add to note 7, after the long paragraph:

Tenniel also darkened the noses of the Duchess (page 60) and the Queen of Hearts (page 82), suggesting that they, too, were boozers.

125. Add to note 8:

Critics have observed that the first card to fall is the Ace of Clubs, the executioner.

136. Add to note 2:

Many attempts have been made to justify the eccentric moves in Carroll's chess game. See "Looking-Glass Chess" by Rev. J. Lloyd Davies in the *Anglo-Welsh Review* 19 (Autumn 1970) and "Looking-Glass Chess" by Ivor Davies in *Jabberwocky* (Autumn 1971). The most detailed analysis is A. S. M. Dickins's lecture "Alice in Fairyland," as edited and expanded in *Jabberwocky* (Winter 1976). The Fairyland is "fairy chess," a common term for variants of chess based on unorthodox pieces and rules. The article is cited again in Chapter 3, Note 1, and Chapter 9, Note 1. Incidentally, Bird's opening, mentioned earlier in this note, is pawn to king bishop's four.

137. Put a [0] after the chapter title. Add new note:

0. An early draft of the table of contents for *Through the Looking-Glass*, located in the Houghton Library at Harvard University, shows that this chapter was originally to be called "The Glass Curtain"; Chapter V was once two chapters, "Living Backwards" and "Scented Rushes," most likely divided before the paragraph beginning "She looked at the Queen," (p. 200); Chapter VIII was called "Check!"; and Chapter XII was written in as "Whose Dream Was It?". See Matt Demakos's "The Annotated Wasp," *Knight Letter* 72 (Winter 2003).

146. Add to note 9:

For some strange reason, not yet understood, in this picture and the next Tenniel gave the Red King the same crown as worn by queens! Was it simply a blunder on his part? If so, why did Carroll, who surely knew that chess kings are topped with a cross, not object?

152. Add to note 18:

In the illustration on page 214, Tenniel pictures the toves with noses that are long helices, like corkscrews. In keeping with the book's mirror symmetry motif, helices come in two forms, each a mirror reflection of the other.

155. Add the following paragraphs to note 42:

See also *Knight Letter* 70 (Winter 2002). The issue features a lengthy discussion of foreign translations of "Jabberwocky" with an abundance of examples. Written by August Imholtz, the article first appeared in *The Rocky Mountain Review of Language and Literature* (Vol. 41, No.4, 1997).

Jabberland, a Whiffle Through the Tulgey Wood of Jabberwocky Imitations is a collection of more than 200 parodies of "Jabberwocky"! It was printed in 2002, edited by Dayna McCausland and the late Hilda Bohem. Copyright laws prevented the book from going on sale, but a limited edition was offered to members of the Lewis Carroll Societies in the United States and Canada.

It is easy to write nonsense parodies of "Jabberwocky": simply substitute new nonsense words for Carroll's. More difficult is to substitute words that make a sensible lyric poem. For example, Harvard professor Harry Levin, in his fine essay "Wonderland Revisited" in *Jabberwocky* (Autumn 1970), does just this to produce the following lovely quatrain:

'Twas April and the heavy rains
Did drip and drizzle on the road:
All misty were the windowpanes,
And the drainpipes overflowed.

162. Place a [9a] for another annotation at end of third paragraph from the bottom. Add an annotation:

9a. Mathematician Solomon Golomb commented in a letter on the Red Queen's remark: "When the Red Queen says, 'When you say "hill," *I* could show you hills, in comparison with which you'd call that a valley,' and Alice objects, 'a hill can't be a valley, you know. That would be nonsense,' I suspect that Dodgson was reacting to something in Hans Christian Andersen's story '*Elverhøj*' (*Elf Hill*, which is very famous and was even made into a ballet). The Troll King (the Mountain King, or *Dovreguben* in Ibsen's *Peer Gynt*, written later) from Norway, is visiting the Elf King in Denmark; and the Troll King's ill-mannered son says, regarding the 'Elf Hill' of the title, 'You call this a *hill*? In Norway, we would call it a *hole*!' (Denmark is very flat and Norway is very mountainous.) Alice expresses Dodgson's mathematical view that what is convex cannot be concave. (We would need to know when the English translation of "*Elverhøj*" reached Oxford, and if Dodgson is likely to have read it.)"

181. Add to note 1:

A much later jingle is worth quoting:

A divinity student named Tweedle
Refused to accept his degree.
"It's bad enough to be Tweedle," he said,
"Without being Tweedle, D.D."

Note that "Fiddle" can be substituted for "Tweedle."

202. Add to note 13:
Solomon Golomb writes: "Many of your readers would be surprised to learn that this is precisely the kind of spinning four-sided top called a 'dreidle,' with which Jewish children play on Chanukah. The four Hebrew letters ה, ג, נ, and ש, are on the four sides, instructing the player, respectively, to take a) nothing, b) everything, c) half, or d) add something to the pot."

208. Put [4a] at the end of the first line below the poem. Add new note:
4a. Alice's version of the nursery rhyme, with its faulty last line, actually appeared in an 1843 London book titled *Pictorial Humpty Dumpty*. See Brian Riddle's "Musings on Humpty Dumpty," in *Jabberwocky* (Summer/Autumn 1989). The jingle's final line is usually "Couldn't put Humpty together again."

223. Add to note 4.
James Tertius DeKay and Solomon Golomb each wrote to suggest that Haigha and Hatta may have been suggested by the names of two fifth-century brothers, Hengist and Horsa. The early Saxons traced their lineage back to these two warriors.

226. Add to note 8:
See "Carroll, the Lion and the Unicorn," by Jeffrey Stern, in *The Carrollian* (Spring, 2000).

241. Add to the end of the fourth paragraph from the bottom (after "deep ditch") the numeral [8a]. Add new note:
8a. Frankie Morris, in *Jabberwocky* (Autumn, 1985) conjectures that the White Knight's inability to stay on his horse may have reflected the notoriously bad horsemanship of King James I. Sir Walter Scott's novel *The Fortunes of Nigel* said the king had a specially constructed saddle to keep him locked on his horse, and Dickens, in his *Child's History of England*, called James I "the worst horse rider ever seen." In 1692 his horse stumbled and threw him into an icy river so that only

his boots were visible. This may have inspired Tenniel's drawing of Alice rescuing the White Knight from a ditch.

246. Add the following paragraph between the first two paragraphs of the note:
All the stanzas of Wordsworth's poem were printed in the first edition (1960) of *The Annotated Alice*. They were omitted from this edition for space reasons.

247. Add to note 15:
Leslie Klinger, in the first volume of his *New Annotated Sherlock Holmes* (Norton, 2004), reproduces (page 428) an advertisement for Rowland's macassar oil. Klinger writes in a note that the oil was made from ylang-ylang, a perfume extracted from a tropical Asian tree. He adds that the name *macassar* derives from Makasar, an Indonesian city now called Ujung Pandang.

248. Add to note 18:
The present steel suspension bridge was built during 1938–46. See Ivor Wynne Jones's article "Menai Bridge" in *Bandersnatch* 127 (April 2005).

260. At the end of the third paragraph from the bottom, after "thirty-times-three," add the numeral [13a]. Add new note:
13a. Three-times-three was and still is a popular way of ending a toast with 3 × 3 = 9 cheers. Tennyson, in the conclusion of "In Memoriam," writes:

> Again the feast, the speech, the glee . . .
> The crowning cup, the three-times-three.

263. Add to note 15:
Solomon Golomb wrote to say that the British word "pudding" is much more vague than as used here. "It is any sort of sweet or dessert, or even a different food entirely, as in Yorkshire pudding." Note that the pudding invented by the White Knight (page 242) was intended for the "meat course."

274. Add to note 1:
In the last two lines of the first stanza, Carroll rhymes *sky* with *dreamily*. According to R. J. Carter, in a note to his fantasy *Alice's Journey Beyond the Moon* (Telos, 2004), Dean Liddell pronounced *university* to rhyme

with *sky*. The same rhyming occurs in the following jingle, which Carter quotes. It was often recited at Oxford University in Carroll's time.

I'm the Dean of Christ Church;—Sir
There's my wife, look well at her.
She's the Broad and I'm the High:
We're the University.

281. Add in the margin:
In 2005, at a Christie's auction in New York City, the galleys sold for $50,000.

298. Add a postscript:
In 1978, England's Lewis Carroll Society sponsored a symposium at which Carroll scholars debated at length the question of whether galleys of the "Wasp in a Wig" episode were authentic or an impressive forgery, as detailed in *Jabberwocky* (Summer 1978). Arguments pro and con were given, but the majority opinion was on the side of authenticity. However, all agreed that in the episode, although it presented Alice in a new light, the writing was not Carroll at his best, and that Tenniel was justified in suggesting that it be excluded from the book. Also debated was whether the episode was intended as a chapter, or as part of the chapter about the White Knight. There was no agreement on where Carroll intended the episode to be placed.

For a comprehensive account of the symposium, including the reproduction of newly discovered documents bearing on the questions, see Matthew Demakos's article "The Authentic Wasp" in *Knight Letter* 72 (Winter 2003).

308. Add to the end of the bibliography:
Artist of Wonderland: The Life, Political Cartoons, and Illustrations of Tenniel. Frankie Morris, 2005.

1 Morton Cohen, *Lewis Carroll: A Biography* (New York: Knopf, 1995).

2 Martin Gardner, *The Universe in a Handkerchief* (New York: Copernicus, 1996); Martin Gardner, *The Annotated Snark* (New York: Simon and Schuster, 1962); Lewis Carroll, *Phantasmagoria,* with introduction and notes by Martin Gardner (Amherst, NY: Prometheus Books, 1998); Lewis Carroll, *The Nursery "Alice,"* with an introduction by Martin Gardner (New York: McGraw Hill, [1966]; Lewis Carroll, *Alice's Adventures Under Ground,* with an introduction by Martin Gardner (New York: Dover, 1965); *Sylvie and Bruno*, with an introduction by Martin Gardner (New York: Dover, 1988).

3 *Looking for Lewis Carroll*, http://carrollmyth.com/.

4 Karoline Leach, *In The Shadow of the Dreamchild: A New Understanding of Lewis Carroll* (London: Peter Owen, 1999), 71, 223.

5 Ibid., 196, 252–56.

6 Dan Brown, *The Da Vinci Code* (New York: Doubleday, 2003).

7 Morton Cohen, "When Love Was Young: Faliled Apologists for the Sexuality of Lewis Carroll, *The Times Literary Supplement,* September 10, 2004: 12–13.

8 Karoline Leach, "Ina In Wonderland" and "The Real Scandal: Lewis Carroll's Friendships with Adult Women," *The Times Literary Supplement*, August 20, 1999, and February 9, 2002.

9 Karoline Leach, "'Lewis Carroll' as Romantic Hero: Anne Thackeray's *From an Island*," *The Carrollian* 12 (Autumn 2003): 3–21.

10 [*Hexham is not actually stated as being from Christ College, Cambridge. At the end of the novel, a friend merely writes from "Ch. Coll., Cambridge" to Hexham. Earlier on, Hexham writes from Lyndhurst, the novella giving no stated connection to the town. – Matt Demakos*]

11 Keith Wright, letter to the editor, *The Carrollian* 13 (Spring 2004): 59–60.

12 Matthew Demakos, letter to the editor, *The Carrollian* 14 (Autumn 2004): 63–64.

13 Linda Sunshine, *All About Alice* (New York: Clarkson Potter, 2004).

14 Linda Sunshine, *All About Oz* (New York: Clarkson Potter, 2003).

15 Martin Gardner, *On the Wild Side* (Amherst, New York: Prometheus Books, 1992).

FESTSCHRIFT

A Triad of Mathematics Popularizers: Martin Gardner, Richard Proctor, & Charles Dodgson

FRANCINE F. ABELES

Martin Gardner (1914–2010), Richard A. Proctor (1837–1888), and Charles L. Dodgson (1832–1898) "met" in the summer of 1993 on the manuscript pages of *The Mathematical Pamphlets of Charles L. Dodgson and Related Pieces*, which I was editing, when I received a letter from Martin about some pieces in the popular nineteenth-century British science journal, *Knowledge*, that had been found by the British mathematician and consummate puzzle expert, David Singmaster. Martin had been sending me relevant reference articles as well as his opinion of drafts of sections of the book.

I met Martin and David at the first "Gathering for Gardner" (G4G1) in 1993, and saw Martin again in 1996 at G4G2, the only other one of these meetings he attended. This event, held in Atlanta, Georgia in even-numbered years beginning in 1996, last occurred in 2010, the ninth such meeting (G4G9). They attract mathematicians deeply interested in games

Francine F. Abeles holds a joint appointment as Professor in the Department of Mathematics and in the Department of Computer Science at Kean University in Union, NJ. Her primary research interest is in the history of mathematics and computing in the 19th and early 20th centuries. She edited the second volume of the series The Pamphlets of Lewis Carroll, *which focuses on his mathematical work, as well as the fourth, on his logic. Forthcoming in 2011 are papers on quasideterminants, Dodgson's view of non-Euclidean geometry, and a joint work with A. Moktefi on crosscurrents in geometry and logic between Dodgson and H. MacColl.*

and puzzles, as well as those who construct these ingenious puzzles. Elwyn Berlecamp, John Conway, Donald Knuth, and Raymond Smullyan are some of the illustrious attendees.

The story behind that virtual meeting is one of the three aims of this article. Another is a closer look at Proctor, Gardner's nineteenth-century counterpart, from the viewpoint of the mathematical pieces he wrote in *Knowledge*. The major source of information on Proctor and the journal he founded is Bernard Lightman's exhaustive study " 'Knowledge' Confronts Nature: Richard Proctor and Popular Science Periodicals," but there is no discussion of his mathematical articles, so Proctor's efforts to popularize topics in mathematics are unknown. The third is a description of Dodgson as a popularizer of mathematical and logic(al) notions. We take these aims in reverse order.

CHARLES DODGSON'S ATTEMPTS TO POPULARIZE MATHEMATICS AND LOGIC

Dodgson was both a popularizer and an educator of mathematics, especially of logic. While he was the Mathematical Lecturer at Christ Church, he often gave free private instruction to family groups of parents, their children, and their children's friends in their private homes on such mathematical topics as ciphers, particularly his *Memoria Technica* cipher, arithmetical and algebraical puzzles, and an algorithmic method to find the day of the week for any given date. He originally created the *Memoria Technica* cipher in 1875 to calculate logarithms, but found many more uses for it as a general aid to remembering, writing a simplified version of it for teaching purposes in 1888.

The topics he chose to teach privately focused on memory aids, number tricks, computational shortcuts, and problems suited to rapid mental calculation, developing this last topic into a book, *Curiosa Mathematica, Part II: Pillow Problems*, published in 1893. He continued to provide instruction in this way on logic topics. He also gave logic lessons in his rooms at Christ Church. In June 1886 he gave lectures at Lady Margaret Hall, Oxford and in May 1887 at the Oxford High School for Girls. He used material that he eventually incorporated into his book *The Game of Logic*, a work he had essentially completed in July 1886, but that did not appear until November in an edition Dodgson rejected for being substandard. The second (published) edition came out in February of the following year. Dodgson hoped the book would appeal to young people as an amusing mental recreation. The objective of the "game," played with a board and counters, is to solve syllogisms (roughly, an argument having two premises and a single conclusion). He found this book, and even more so, his *Symbolic Logic, Part I: Elementary* essential in teaching students. He believed his own book on symbolic logic was far superior to those in current use. One of the major reasons he chose to write it under his pseudonym, Lewis Carroll, is that he wanted the material to appeal to a large general audience, particularly to young people, a task made easier using the wide acclaim accorded him as the writer of the *Alice* books.

On August 21, 1894, answering a letter from a former child friend, Mary Brown, now aged thirty-two, he wrote,

> You ask what books I have done. . . . At present I'm hard at work (and have been for months) on my Logic-book. (It really has been on hand for a dozen years: the "months" refer to preparing for the Press.) It is Symbolic Logic, in 3 Parts—and Part I is to be easy enough for boys and girls of (say) 12 or 14. I greatly hope it will get into High Schools, etc. I've been teaching it at Oxford to a class of girls at the High School, another class of the mistresses (!), and another class of girls at one of the Ladies' Colleges. (Cohen 1979. 1031)

In a letter dated November 25, 1894 to his sister Elizabeth, he wrote,

> One great use of the study of Logic (which I am doing my best to popularise) would be to help people who have religious difficulties to deal with, by making them see the absolute necessity of having clear definitions, so that, before entering on the discussion of any of these puzzling matters, they may have a clear idea what it is they are talking about. (Cohen 1979, 1041)

He believed that mental activities and mental recreations, like games and particularly puzzles, were enjoyable and conferred a sense of power to those who make the effort to solve them. Earlier, in 1885, Carroll had published *A Tangled Tale*, a book whose chapters he called "knots." There were ten of them and they contained mathematical problems and puzzles that had appeared between April, 1880 and March, 1885 in a magazine titled *The*

Monthly Packet. Readers gave their solutions to the problems and Carroll commented on them.

In an advertisement to promote the sale of *Symbolic Logic, Part I*, he wrote,

> I claim, for Symbolic Logic, a very high place among recreations that have the nature of games or puzzles. . . . Symbolic Logic has one unique feature, as compared with games and puzzles, which entitles it, I hold, to rank above them all. . . . The accomplished Logician . . . may apply his skill to any and every subject of human thought: in every one of them it will help him to get clear ideas, to make orderly arrangement of his knowledge, and more important than all, to detect and unravel the fallacies he will meet with in every subject he may interest himself in. (Abeles 2010, 91–92)

The statements of almost all the problems in both parts of his symbolic logic books are amusing to read. This attribute derives from the announced purpose of the books, to popularize the subject. But Dodgson naturally incorporated humor into much of his serious mathematical writing, infusing this work with the mark of his literary genius. In Book (chapter) XII of part two of *Symbolic Logic*, instead of just exhibiting the solution tree piecemeal for a particular problem he gives a "soliloquy" as he works it through, accompanied by amusing "stage directions" showing what he is doing to enable the reader to construct the tree for himself.

MARTIN GARDNER AND RICHARD PROCTOR

Gardner and Proctor, who were born three-fourths of a century apart, had many common interests and talents. Both admired the mathematical side of Carroll/Dodgson, Gardner more for his games and puzzles; Proctor more for his serious mathematical ideas. As writers, both popularized science and mathematics, and both had a gift for lucid written exposition. Gardner enjoyed chess and playing music on a saw. Proctor was proficient in both chess and whist, and played the pianoforte. Gardner published more than 70 books; Proctor more than 60. As the founder and editor ("conductor") of the popular British science journal *Knowledge: An Illustrated Magazine of Science Plainly Worded—Exactly Described* from 1881 until his death, Proctor wrote regularly on his particular specialty, astronomy, on mathematics, and on the relationship between science and religion. Contributors, usually dilettantes, submitted items on a variety of scientific topics that Proctor commented upon, which served as an exchange of ideas on pieces published in the journal. For twenty-five years, from December 1956 to December 1981, Gardner edited the Mathematical Games column of the popular science magazine *Scientific American*, where readers would send in subjects for articles, as well as a great variety of problems and their solutions. (In 1995, David Singmaster privately published an index to topics in the column and the seventeen books Gardner wrote that derived from them.)

PROCTOR'S MATHEMATICAL CONTRIBUTIONS TO *KNOWLEDGE*

Proctor founded *Knowledge* as a weekly journal in November of 1881. From 1882 on, two volumes were published yearly. Some of his articles, book reviews, and replies to correspondents and to contributors of problems and articles he signed with his name; some he signed as editor, some he wrote with pseudonyms, and some he left unsigned. His first mathematical piece, "Betting and Mathematics," appeared in the December 16 issue. Six months later, a follow-up, "Winning Wages," appeared in the June 2 issue. Then Proctor began a long and ambitious series that he called "Easy Lessons." He chose as his topic differential calculus, beginning it on June 30, 1882, continuing it in each issue through the end of the year; returning briefly to it on March 16 in volume III, and ending it in the April 27 issue, where he wrote his only lesson in this series on integral calculus.

On June 1, 1883 he began a second long and equally ambitious series that he called "Geometrical Problems," which appeared in every issue through August 31, 1883. Also, a three-part series, "Notes on Euclid," appeared in all of the January 1883 issues. Two issues in February carried his pieces on compound proportion, and in "A Mathematical Letter" in the first issue of March, Proctor reminisced about his Cambridge days. In the August 31 issue, he provided a solution to a "Singular Numerical Property" problem. His piece "Easy Lessons in Geometrical Problems" appeared in the September 14 issue, and was followed on November 9 by "Mathematics of the Imaginary," in which Proctor ridiculed Arthur Cayley's presidential address to the British Association for the Advancement of Science that included Lobachevsky's non-Euclidean geometry as well as the possibility of quadridimensional space. He wrote, "But the world is not altogether wanting in common sense, not ready to sing in chorus about the professors of the unintelligible . . . . " (Proctor 1883, 287) His last paper on Geometrical Problems, "Hints on the Solution of Geometrical Problems" appeared in the January 11, 1884 issue. The issue of January 25 carried his piece "Achilles and the Tortoise." On March 7 Proctor wrote "Notes on Euclid's First Book," announcing the beginning of yet another long series of pieces that ran through mid-August, 1884. He also used the title "Easy Riders on Euclid's First Book," and alternated these with "Easy Lessons in Co-ordinate Geometry." Having completed "Notes" on Euclid's First Book early in July, he began Euclid's Second Book in his "Notes" series. Beginning on October 24, 1884 he began another long series, "Chats About Geometrical Measurement," which ended in August, 1885. These pieces dealt with a wide range of topics: lengths of curves, areas of surfaces bounded by curves, volumes and surfaces of solid figures bounded by curved surfaces, the conic sections (parabola, ellipse, hyperbola), volumes and surfaces of solids (cylinder, cone, sphere).

Proctor published nothing on mathematics in volume IX (November 1885–October 1886) when *Knowledge* became a monthly journal, now called volume 1 of the new series, nor in volume II of this new series. In volume III (November 1887–October 1888) there were seven pieces by Proctor. The first, a response to a correspondent, "Note on Euclid (I.32)" on November 1; the second, in the July 2 issue, "Easy Study in the Differential Calculus," was his response to an article published in the *Saturday Review* concerning his two books *First Steps in Geometry* and *Easy Lessons in the Differential Calculus*, which were derived from Proctor's pieces in *Knowledge*. In the same issue he wrote two items, "Bacon's Own Cipher" followed by "An Undecipherable Cipher." "The Donnelly Cipher Deciphered" appeared in the next issue and in the following issue he published "Mr. Donnelly's Ciphering." In this same issue of September 1, the final issue in which he wrote a mathematical piece, Proctor authored "Easy Problems in the Differential Calculus" where he dealt with some problems suggested by his piece in the earlier issue of July 2.

DODGSON AND PROCTOR ON THE PAGES OF *KNOWLEDGE*

Dodgson wrote two pieces that were published in *Knowledge*. The first was a response to the inaugural article in Proctor's series, "Chats

About Geometrical Measurement," this one in the form of a dialogue between A. and M. that was published in the issue of October 24, 1884. Two weeks later, on November 7, Dodgson's short piece titled "Euclid's Theory of Parallels" appeared. His response dealt with points Proctor had made that Dodgson faulted from a logic point of view. Proctor replied to Dodgson, refuting his objections, in an article, "Euclid's Theory of Parallels," which was published in the issue of December 26. But from a review of Dodgson's pamphlet *The Principles of Parliamentary Representation* in the November 28 issue we see that Proctor, who was most likely the reviewer, greatly admired both the pamphlet and its author from a mathematical point of view. He wrote, "[W]e must afford unstinted praise to Mr. Dodgson's tract, which lays down, on mathematical principles, the numbers of voters to be assigned to each electoral district, and the number of members to be returned." (Proctor 1884, 449)

The second of Dodgson's pieces was a response to a problem submitted by H. Askew that appeared in *Knowledge* on May 30, 1884. Dodgson's response, "Divisibility by Seven," was published on July 4, 1884. It is reproduced in Abeles 1994, 277–79.

ENTER MARTIN GARDNER

In my assembling the mathematical pamphlets and ephemera for *The Mathematical Pamphlets*, "Divisibility by Seven" was one of the most difficult items about which to obtain accurate information. It is not listed in *The Lewis Carroll Handbook* (*LCH*), the standard reference to Carroll's work, and its citation and date of publication are both incorrect in *Lewis Carroll at Texas*, the catalog of the well-regarded Warren Weaver collection at the Harry Ransom Humanities Research Center (HRHRC) at the University of Texas in Austin. The proof sheet of "Divisibility by Seven" in the Morris L. Parrish collection at Princeton University has the signature of Falconer Madan, author of the first edition of the *LCH*, with his comments, "Not Mathematical Questions (nor Educational Times), I think. Not? English Mechanic."[1] With the assistance of the staff of the British Library, we checked each of these publications for the most likely period of publication, 1881–97. Madan was right! Then with the help of the HRHRC staff, we checked another dozen periodicals in the same time period. Nothing turned up. Perhaps the item had never got beyond the proof-sheet stage.

In the summer of 1993, a letter arrived from Martin, dated July 28, with copies of some pieces on geometry by Dodgson in *Knowledge* that had been found by David Singmaster. In the "Letters to the Editor" sections of these pages I recognized that the font style, but not the size, and the square brackets were exactly the same as those in the proof sheet of "Divisibility by Seven." Assuming the font size was probably a function of the number of Letters to the Editor to be included in a particular issue, I wrote to friends in London asking that they check the relevant section of *Knowledge* from May 30, 1883 through 1885.[2] Sure enough, Askew's letter had appeared on May 30, 1884; Dodgson's response, number 1324 on p. 15 of volume VI, had been published on July 4, 1884.

Puzzles that appear difficult can have easy solutions if you know how to approach them.

Had I approached this identification problem as a puzzle, I might have chosen *Knowledge* as a journal to search because Dodgson wrote in his diary entry of June 10, 1884 that he had "sent a corrected proof of my letter on Divisibility by 17 to Knowledge." But he meant 7, not 17, a transcription error on p. 426 of Roger L. Green's edition of *The Diaries of Lewis Carroll* not easily recognized as such because Carroll had written in his diary just a week earlier that he had invented rules for dividing numbers by 17 and 19.

As an example of Dodgson's rule, consider testing the number N= 8,026,518,423 for divisibility by 7. First mark off N from the right in periods of three: $423 + 518(1000) + 026(1000)^2 + 8(1000)^3$. Then take the difference between the sums of the alternate periods: $423 - 518 + 026 - 8 = -77$, and call the result M. If M is divisible by 1001 or any of its factors—7, 11, 13—so is N. (If the periods are single digits or two digits in length we have a test for divisibility by 11 and 101, respectively.) How valuable is Dodgson's rule? The question is left for the reader to decide.

CONCLUSION

Gardner, Proctor, and Dodgson popularized mathematics. But their motives and aspirations were quite different. For Dodgson, mathematics, and logic particularly, were essential to the fabric of his life. As a gifted written communicator, he turned to explicit popularization on a broad scale in the final decade of his life.

As the editor of the *Monthly Notices of the Royal Astronomical Society*, a writer of articles for many different periodicals, and the author of popular science books, Proctor had founded *Knowledge* to directly compete with the well-regarded periodical *Nature* for control of the periodical market aimed at the rapidly growing reading public. No sections or articles intended for professional mathematicians were included. Critical of professional scientists, Proctor's goal was to bring current science and mathematics to the attention of the public.

Before he began writing for *Scientific American* at the age of forty-two, Gardner had been a freelance writer and newspaper reporter whose lack of a formal mathematical education forced him to focus on problems, puzzles, paradoxes, and mathematical topics that would appeal to a sophisticated, but not necessarily mathematically trained (although mathematicians, too, were attracted) audience. Gardner expressed his affection for Dodgson's writings in three annotated editions of the *Alice* books, two editions of the annotated *Snark*, and mathematical topics connected with Dodgson that he included in his *Scientific American* column. In his book *The Universe in a Handkerchief*, on Carroll's mathematical games, puzzles, and wordplays, Gardner wrote, "[W]hile writing a column on recreational mathematics for *Scientific American*, I discovered that Carroll not only shared my enthusiasm for play mathematics . . . he also shared my hobby of conjuring. The more I learned about the life and opinions, the more I came to feel a spiritual kinship with him." (Gardner 1996, *ix*)

Martin Gardner, Richard Proctor, and Charles Dodgson were pioneers in popularizing mathematics. Their work has provided the foundation for the growing body of scientific knowledge currently available to the public in written and electronic formats.

[1] The *Educational Times*, a monthly London periodical, began publishing several columns on mathematical problems and their solutions in 1862. These were reprinted annually in two volumes as *Mathematical Questions and Solutions, from the "Educational Times," with Many Papers and Solutions in Addition to Those Published in the "Educational Times."* Dodgson contributed eleven items on various topics, but contributed nothing to the periodical the *English Mechanic*.

[2] H. Askew's letter, to which Dodgson had responded in "Divisibility by Seven," bore the incomplete date May 30 and the number 1274. Singmaster's material contained a sufficient number of dates of Letters to the Editor to permit a narrowing of the period to check for both Askew's letter and Dodgson's reply.

References

Abeles, Francine F., ed. *The Mathematical Pamphlets of Charles Lutwidge Dodgson and Related Pieces*. New York: LCSNA, 1994.

_______. The *Memoria Technica* Cipher. *Cryptologia* XXVII July 2003, 217–229.

_______., ed. *The Logic Pamphlets of Charles Lutwidge Dodgson and Related Pieces*. New York: LCSNA, 2010.

Carroll, Lewis. *A Tangled Tale*. New York: Dover, 1958 [1885].

Cohen, Morton. N., ed. *The Letters of Lewis Carroll*. Two volumes. New York: Oxford University Press, 1979.

Dodgson, Charles L. Divisibility by Seven. *Knowledge* VI, July 4, 1884, 15.

________. Euclid's Theory of Parallels. *Knowledge* VI, Nov. 7, 1884, 390–91.

________. *Curiosa Mathematica, Part II: Pillow Problems*. New York: Dover, 1958 [1893].

Gardner, Martin. *The Universe in a Handkerchief. Lewis Carroll's Mathematical Recreations, Games, Puzzles, and Word Plays*. New York: Copernicus, 1996.

_______. *Logic Machines and Diagrams*. New York: McGraw Hill, 1958.

Green, Roger L., ed. *The Diaries of Lewis Carroll*. Two volumes. New York: Oxford University Press, 1954.

Lightman, Bernard. "Knowledge" Confronts "Nature": Richard Proctor and Popular Science Periodicals. In Louise Henson, et al. eds. *Culture and Science in the Nineteenth Century Media*. Aldershot, Hants, England: Ashgate Publishing, 2004, 199–210.

Proctor, Richard A. Mathematics of the Imaginary. *Knowledge* IV, Nov. 9, 1883, 287–288.

_______. Chats About Geometrical Measurement. *Knowledge* VI, Oct. 24, 1884, 334-339.

_______. Review of *The Principles of Parliamentary Representation*. By Chas. L. Dodgson, M.A. *Knowledge* VI, Nov. 28, 1884, 449.

Singmaster, David. *Index to Martin Gardner's Columns and Books*. Tech. Rpt. SBU-CISM-95-09. London: South Bank University, 19 August 1995.

One More Class: Martin Gardner & Logic Diagrams

AMIROUCHE MOKTEFI & A. W. F. EDWARDS

Martin Gardner (1914 – 2010) was a multifaceted author; one of his lifelong interests was in logic diagrams and machines. In 1958, he gave a pioneering account of early approaches to creating tables and diagrams for logic in the first edition of his *Logic Machines and Diagrams*. He followed it with an encyclopedia article in 1967 and a second edition of the book in 1983. The book became an "unforgettable classic" (Donald Michie's description in his Foreword to the second edition) that revived the interest in logic diagrams and has influenced scholars ever since.

Amirouche Moktefi is a postdoctoral fellow at the LHSP Archives Poincaré laboratory (Nancy University). He is also an associate member of the IRIST laboratory (University of Strasbourg). His area of research includes the history of mathematics and logic. He obtained his Ph.D. in 2007 with a dissertation on Lewis Carroll's symbolic logic, a survey of which was published as: "Lewis Carroll's Logic" in D. M. Gabbay and J. Woods (eds.), British Logic in the Nineteenth Century, *vol. 4 of the* Handbook of the History of Logic *(Amsterdam: North-Holland, 2008), pp. 457–505.*

A.W. F. Edwards is a Fellow of Gonville and Caius College in Cambridge, and retired Professor of Biometry in the University. His experience teaching the binary logic of Mendelian genetics combined with an interest in the life of John Venn, who spent his whole career in the College, led him to study Venn diagrams and to write Cogwheels of the Mind: The Story of Venn Diagrams *(John Hopkins University Press, 2004). His other books include* Likelihood *(Cambridge University Press, 1972),* Foundations of Mathematical Genetics *(Cambridge University Press, 1977) and* Pascal's Arithmetical Triangle *(John Hopkins University Press, 1987).*

In addition to initiating and making several generations of readers enjoy the field, Gardner himself also contributed to its development. Indeed, in his 1958 book he explains how a particular type of diagram, known as Venn diagrams, could be used advantageously for propositional calculus and not just to class logic, as they had usually been employed until then. One has to remember that at the time Gardner wrote his book, the topic of logic diagrams was almost forgotten, since the modern logic tradition had excluded visual schemes from proofs, and relegated them to educational contexts. In a way, Gardner's work revived the interest in a field that knew a golden age in the nineteenth century before becoming almost extinct. Fortunately, diagrammatic logic has in the last few decades made important developments, and we now have interesting studies that highly praise the role of diagrams in reasoning [Shin, 1994; Allwein and Barwise, 1996].

CIRCLES AND ELLIPSES

The use of diagrams in logic is old. The most popular method requires the employment of circles to represent classes of objects. There are two variants of this method. The first was popularized by the mathematician Leonhard Euler (1707 – 1783) in his *Letters to a German Princess* (1768). He represents the actual logical relationship between the classes by the topological relations of the circles. For instance, to represent the proposition "all x are y," one has just to draw a circle x inside a circle y [Fig. 1]. This looks very simple and intuitive but a close look shows that it contains some ambiguity. Indeed, we do not know for sure whether the class x should be *strictly* included within the class y, since x might be identical to y without violating the information contained in the above proposition.

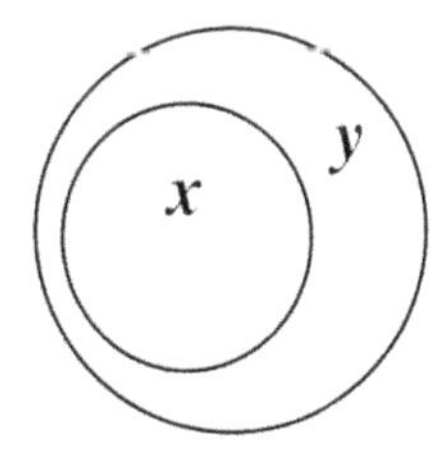

Fig. 1

To avoid this ambiguity, in 1880 the logician John Venn (1834 – 1923) published a second scheme in which he proceeds as follows. First he draws figures (e.g., circles) which intersect each other in such a way as to divide the space into 2^n compartments corresponding to the 2^n combinations between the n classes involved in the argument. For instance, if we have 2 terms, x and y, we have 4 combinations (and compartments): *x y*, *x not-y*, *not-x y*, and *not-x not-y*, which can easily be represented by making two circles intersect. So far, the diagram says nothing about the actual relation between *x* and *y*. In order to do so, one has to mark the compartments to indicate their condition. For instance, shading a compartment means that it is empty. With the help of this convention, it is easy to represent the proposition "all *x* are *y*" alluded to earlier. Indeed, it says that "no *x* is *not-y*" and hence, that the compartment *x not-y* is empty and should be shaded [Fig. 2].

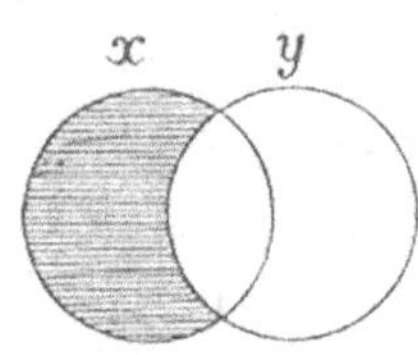

Fig. 2

Venn's method might seem less intuitive than Euler's. Indeed, one has to "read" a Venn diagram in order to "see" the proposition that is represented. However, in addition to their precision, Venn's diagrams surpass Euler's in their capacity to be extended more easily, to deal with further classes in just one diagram. For instance, adding a third term *z* to an Euler diagram might be difficult. Often, one needs more than a single diagram to represent all the possible relations of the term *z* with the other terms *x* and *y*. With Venn however, the extension is very simple at this stage. All one has to do is to draw a circle *z* that intersects the existing circles *x* and *y* in such a way as to divide the existing compartments into two subdivisions: one within *z* and the other outside it [Fig. 3]. Venn's method is a simple application of the dichotomy procedure, where one divides a class *a* into subclasses *a b* and *a not-b*. This method of division was already known to ancient logicians but interest in it had grown with the revival of logic in the second half of the nineteenth century. One finds it, for instance, at the root of William Stanley Jevons' (1835 – 1882) "logical alphabet," put forth in his *Principles of Science* (1874).

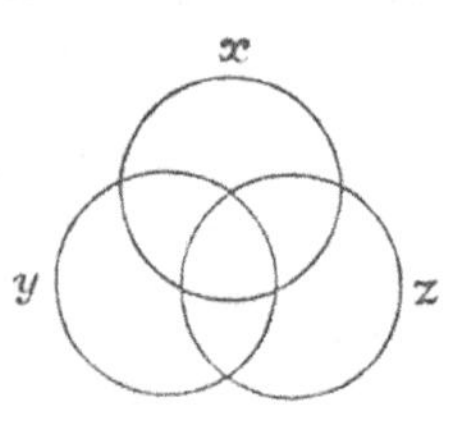

Fig. 3

Venn was a champion of the symbolic treatment of logic that had been initiated by the mathematician George Boole (1815 – 1864) a few decades earlier. Venn considered symbolic logic as a generalization of the old syllogistic, and hence aimed his methods at arguments more complex than mere syllogisms (which require just 3 terms). Consequently, one should not underestimate the "strategic" importance of the extension problem in the work of Venn and his followers, although it was obvious that from, say, 5 or 6 terms onwards, the diagrams would become too complex to provide the visual aid one would look for in such a device. For 4 terms (*x, y, z, w*), Venn explains that it was no longer possible to get the desired diagram with circles, so he replaced them by ellipses [Fig. 4]. The starred compartment corresponds to the subclass *non-x y z w*. It is important to understand here that Venn's abandonment of circles did not result from his incapacity to add a fourth continuous figure to his 3-circle diagram in the desired manner, as he had done at first. Indeed, in his 1880 paper, Venn reproduced such a diagram [Fig. 5]. Venn's preference for ellipses is rather motivated by the simplicity and symmetry of the diagram he obtained [Venn, 1894, p. 116].

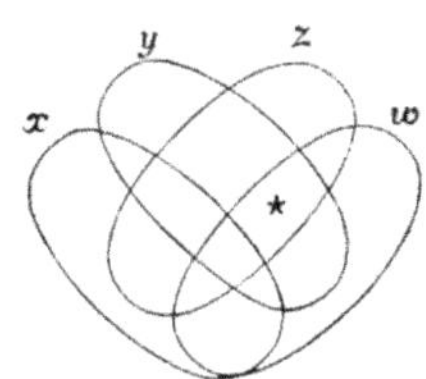

Fig. 4

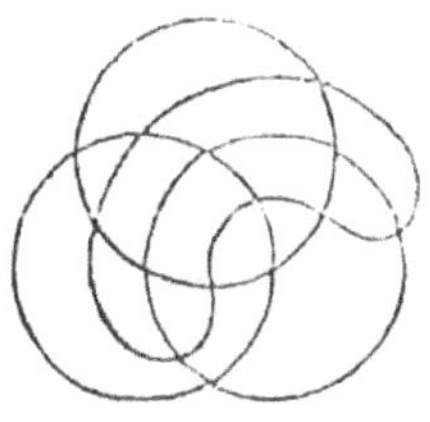
Fig. 5

Naturally, the situation becomes more complicated for 5 terms. Venn failed to draw an appropriate diagram using ellipses. Consequently, for problems requiring 5 terms (*x, y, z, w, v*), he suggested the use of a diagram where one class (*z*) is represented by an annulus (a "doughnut," Gardner calls it), so that subclass *not-z* is represented by a discontinuous space [Fig. 6]. After Venn it was long thought impossible to draw a Venn diagram using 5 ellipses, as Gardner himself mentioned in his "Logic Diagrams" entry [1967, p. 78]. In fact, one has to wait until 1975 to see a 5-ellipse Venn diagram in print [Fig. 7], and others followed in subsequent years [Grünbaum, 1975; Hamburger and Pippert, 2000]. Gardner recorded this discovery in the second edition of his *Logic Machines and Diagrams* [1983, p. 59]. Venn was not happy with his 5-term diagram. However, he preferred using it rather than the asymmetric figure with continuous regions that he knew about and reproduced in his 1880 paper [Fig. 8].

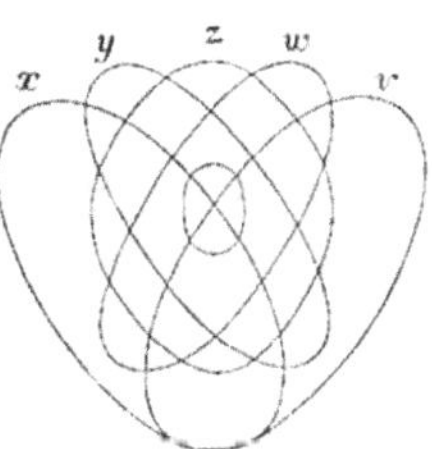

Fig. 6

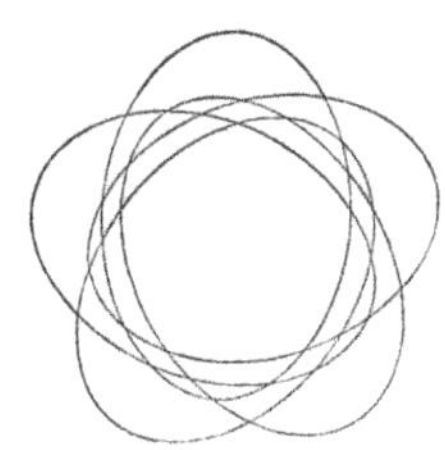
Fig. 7

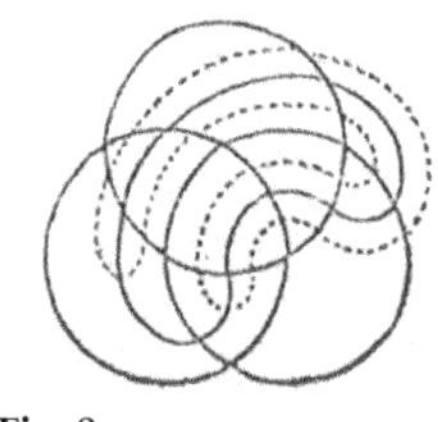

Fig. 8

Venn did not believe it worthwhile to work on diagrams for more than 5 terms. Indeed, he wrote, "Beyond five terms it hardly seems as if diagrams, of the particular kind here described, offer much help, but then we have seldom occasion to trouble ourselves with problems which would introduce more than that number" [Venn, 1894, p. 117]. Venn eventually suggested a solution for 6 terms, which is to put two 5-term diagrams side-by-side, one corresponding to the affirmation of the sixth term and the other to its denial. It is however obvious that such a scheme does not fit the general method previously described, making the diagram lose its visual impact.

SQUARES AND STRIPS

The reader might infer from the exposition above that Venn failed to find a method for drawing a diagram for any number of terms. That is what Lewis Carroll (1832 – 1898) also suggested when he wrote, severely, "Beyond six letters Mr. Venn does not go" [Carroll, 1897, p. 176]. The truth, however, is that Venn admirably solved the problem thanks to an ingenious inductive method which he used to draw [Fig. 5] and [Fig. 8] above, and which could easily be extended to any number of terms. As Venn himself explains:

> It will be found that when we adhere to continuous figures, instead of the discontinuous five-term figure given above, there is a tendency for the resultant outlines thus successively drawn to assume, after the first four or five, a comb-like shape. If we begin by circles or other rounded figures the teeth are curved, if by parallelograms then they are straight. Thus the fifth-term figure will have two teeth, the sixth four, and so on, till the $(4+x)^{th}$ has 2^x teeth. There is no trouble in drawing such a diagram for any number of terms which our paper will find room for. But as has already been repeatedly remarked, the visual aid for which mainly such diagrams exist is soon lost on such a path. [Venn, 1894, p. 118]

Venn's use of discontinuous figures was not motivated by his failure to draw diagrams with continuous figures, but rather by his search for symmetry and simplicity. In brief, Venn was looking for a "nice" diagram. Although it is not fully clear what a nice Venn diagram would be, one can assume that it should be one that provides some visual aid, that it should be easily constructed, and in which compartments are easily identified. All these features would make such a diagram easy to use.

After Venn, several logicians shared his concerns and tried to construct "nice" diagrammatic schemes that would supersede his method, especially when it comes to arguments involving more than 3 terms. Allan Marquand (1853 – 1924), one of Charles Sanders Peirce's (1839 – 1914) students, was one of the first to react to Venn's innovation. The opening of his 1881 paper in *Philosophical Magazine* (the same journal in which Venn published his diagrams just a year before) is explicit about Marquand's aim:

> It is the object of this paper to suggest a mode of constructing logical diagrams, by which they may be indefinitely extended to any number of terms, without losing so rapidly their special function, viz. that of affording visual aid in the solution of problems. [Marquand 1881, 266]

Marquand's scheme is Venn-type in the sense that he first represents all possible combinations of classes before marking the compartments. His method of construction is different, however: Marquand first represents the universe with a square, and then he divides it dichotomously to obtain the compartments, without paying attention to the continuity of the classes. For 4 terms (*A, B, C, D*), one obtains a diagram where the classes *C, not-C, D,* and *not-D* are discontinuous [Fig. 9]. The continuity condition, which Venn kept until his 5-term diagram where he abandoned it with regret, is simply absent in Marquand's requirements. That makes his diagrams more "regular" and easy to draw and to extend, but it looses a good deal of the visual advantage one expects from diagrammatic schemes.

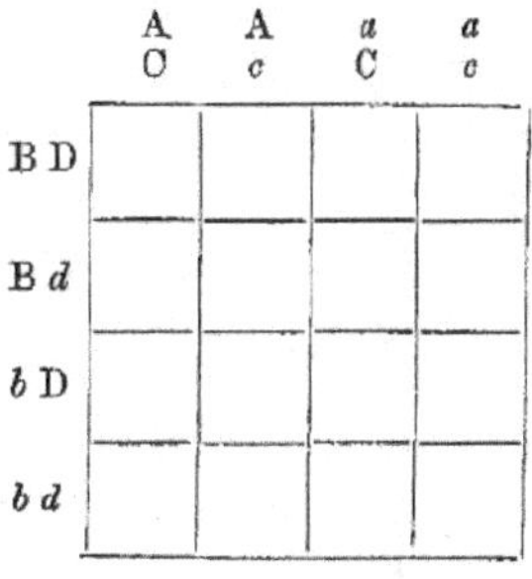

Fig. 9

abcd	*abcd'*	*abc'd*	*abc'd'*	*ab'cd*	*ab'cd'*	*ab'c'd*	*ab'c'd'*		*a'bcd'*	*a'bc'd*	*a'bc'd'*	*a'b'cd*	*a'b'cd'*	*a'b'c'd*	*a'b'c'd'*

Fig. 10

Marquand was not alone in this line. Indeed, Alexander Macfarlane (1851 – 1913) proceeded in the same way when he conceived his own scheme, which he called a "logical spectrum" [Macfarlane, 1885]. In this method, the universe is represented by a "rectangular strip" which is divided horizontally to form the compartments. For 4 terms (*a, b, c, d*) one obtains 16 compartments side by side [Fig. 10]. One wonders whether Macfarlane's spectrum is not just a disguised form of Jevons' logical abacus in which the sequence is now enclosed within a strip. What interested Macfarlane was not the continuity of the figures, but the regularity of the scheme:

> This method allows all the *a* part to be contiguous; but the *b* part is broken up into two portions, the *c* part into four portions, the *d* part into eight portions. However the regularity of the spectrum enables us easily to find all the portions belonging to any one mark. [Macfarlane, 1885, p. 287]

Carroll, to whom we alluded earlier in this paper, also devised a diagrammatic representation for logical problems. His scheme is worth a notice here on two grounds: first because, as is well known, Gardner devoted several papers and books to Carroll's work; and secondly because Carroll's scheme is often presented in secondary literature as a solution to Venn's problem. A close look, however, shows that Carroll, although his scheme differs from Venn's in its construction, proceeds in exactly the same way as Venn when it comes to constructing logic diagrams for more than 4 terms. Carroll published his diagrams for the first time in his *Game of Logic* (1886), where he does not refer to Venn, Marquand, or Macfarlane, and says nothing about constructing diagrams for more than 3 terms. One has to wait for the publication of the first part of his *Symbolic Logic* ten years later (1896) to see his more complex diagrams. Just like Marquand, Carroll represents the universe with a square, which he divides dichotomously to form classes. For 2 terms *x* and *y*, one obtains a diagram with 4 compartments: *xy, x not-y, not-x y, not-x not-y* [Fig. 11]. The addition of a third term *m* is represented by a small square that divides the previous 4 compartments each into 2 subdivisions [Fig. 12].

It is not crucial for the purpose of this paper to discuss how Carroll represented propositions with his diagrams. It suffices to know that like his contemporaries Venn, Marquand, and Macfarlane, he added marks to indicate the state (emptiness or occupation) of the compartments (for more information on this, see [Abeles, 2007] and [Moktefi, 2008]). What interests us here is how Carroll handled the problem of constructing diagrams for more than 3 terms. In order to add a fourth term, Carroll abandoned squares for rectangles [Fig. 13] in a way that recalls how Venn abandoned circles for ellipses.

Adding a fifth term was more difficult. In his diaries, Carroll recorded in 1888 how he added a new figure with "an additional zigzag line working in and out so as to divide each one of the existing 16 compartments" of the 4-term diagram [Wakeling, 2004, p. 434]. Thus he obtains a 5-term diagram where the fifth class is closed and continuous [Fig. 14]. However, in his *Symbolic Logic* (1896), Carroll abandoned this scheme and preferred a new one where the fifth term is added by dividing diagonally each compartment of the 4-term diagram [Fig. 15]. Consequently, the fifth class is discontinuous as it is split into the 16 disconnected subdivisions

x *y*	*x* *non-y*
non-x *y*	*non-x* *non-y*

Fig. 11

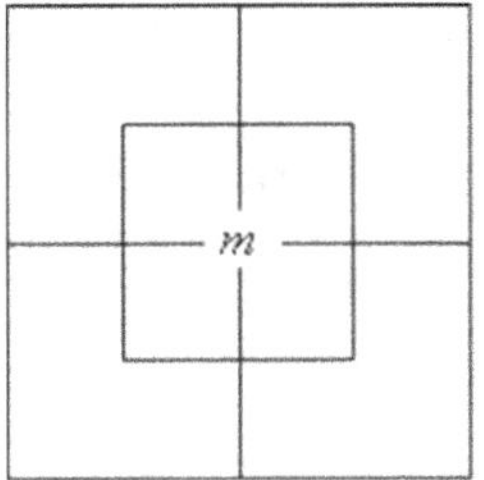

Fig. 12

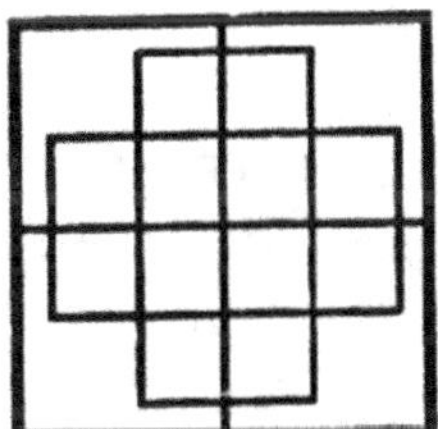

Fig. 13

Fig. 14

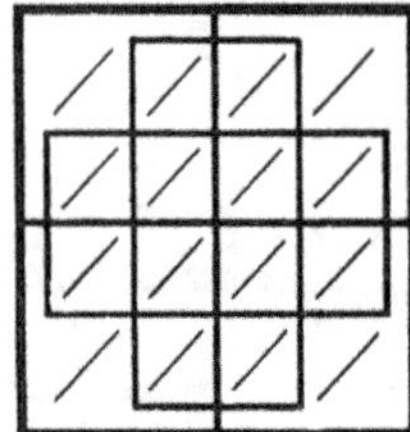

Fig. 15

by the new diagonal lines. Thus Carroll preferred this new scheme, with a discontinuous class, rather than the continuous scheme he knew about and had drawn a few years earlier. He was conscious of the break that happens, as he wrote, "Here, I admit, we lose the advantage of having the *e*-Class all together, 'in a ring-fence', like the other 4 Classes. Still, it is very easy to find; and the operation of erasing it, is as easy as that of erasing any other Class." [Carroll, 1897, p. 177]

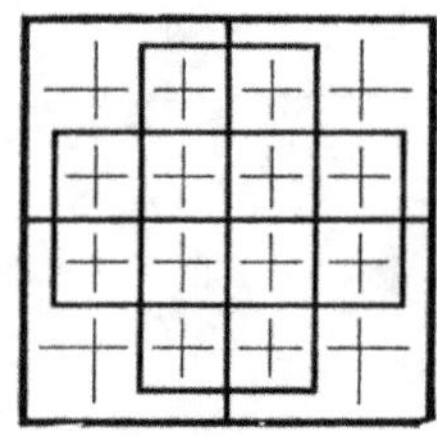

Fig. 16

For more terms, Carroll continued to introduce discontinuous figures. For 6 terms, for instance, he simply inserts a 2-term diagram in each of the 16 compartment of a 4-term diagram [Fig. 16]. It is easy to see how Carroll constructed more complex diagrams similarly. Once the continuity of the figures is no more required, constructing diagrams for *n* terms becomes simple. Such a method of generalization has, for instance, been proposed by Anthony J. Macula in 1995, and one can easily think of others.

FAILURES AND SUCCESSES

A comparison of Venn's and Carroll's methods shows surprising similarities, in spite of the different schemes they employed to construct their diagrams. Indeed, until 4 terms, both Carroll and Venn cared about the continuity of the figures. For the fifth term however, both relaxed that condition and used disjoined regions, in spite of the fact that both knew how to construct (in different ways) a 5-term continuous diagram. Beyond 4 terms, both Carroll and Venn constructed *n*-term diagrams by inserting an *m*-term diagram in each compartment of an *(n-m)*-term diagram. For instance, for $n = 6$, Carroll inserted a 2-term diagram (i.e., $m = 2$) in each of the 16 compartments of a 4-term diagram (see [Fig. 16], above). In order to represent a 6-term diagram (i.e., $n = 6$), Venn, as we have seen earlier, chose to put two 5-term diagrams side by side, with one 5-term diagram corresponding to the assertion of the sixth term and the other corresponding to its denial. Essentially, he had inserted a 5-term diagram (i.e., $m = 5$) in each of the 2 compartments of a 1-term diagram.

It is tempting to suggest that both Venn and Carroll failed in constructing diagrams for any number of terms, since they used discontinuous figures for the fifth term and beyond. One should keep in mind, however, that for logicians, constructing such diagrams was not a mathematical challenge but rather a way to solve logical problems more simply. What interested Venn and Carroll was the efficiency of their diagrams as visual aids rather than their mathematical properties. This is why, as we have seen, both gave precedence to the regularity of their diagrams rather than their continuity. By keeping the diagrams regular, they obtained diagrams that could be easily drawn and used. Carroll's scheme has the advantage of possessing an (apparent) unity and symmetry thanks to the square that encloses the complete diagram, which also makes the division easier. This is why his diagrams are easier to extend, although not necessarily better. Venn himself recognized the superiority of tabular diagrams by using them for problems involving a high number of terms [Venn, 1894, p. 140]. In any case, the use of diagrams for reasoning is hardly advisable when the number of terms increases. Indeed, the figures become too intricate and complex and one quickly loses the visual aid one expects from them. Although it was important for Venn and his contemporaries to show how one could work diagrams for more than 3 terms in order to go beyond traditional syllogistic logic, it was less essential to go beyond, say, 5 or 6 terms, as one seldom meets problems involving more terms.

The issue of constructing logic diagrams for any number of terms remained in this position until mathematicians became interested in the combinatorial problem that hides behind it, that is, how to construct such diagrams without employing disconnected figures at all. In a previous account, the second author (A. W. F. Edwards) sketched clearly how the problem loses its mathematical interest when one relaxes the continuity condition:

> Both Venn and Carroll gave up at four sets and offered five-set diagrams whose fifth set did not consist of a closed curve, so that some regions became disjoint. In our terminology, they were not really Venn diagrams at all: once one admits the possibility of sets being bounded by more than one closed curve, one might as well just list all the binary numbers between 0 and 2^n–1 and put a little ring round each! [Edwards, 2004, p. 32]

As we have already seen, the problem of constructing a Venn diagram with continuous figures for any number of terms had already been solved by Venn himself with his inductive method, even if he did not give a formal proof. Such proofs have been published occasionally in the mathematical literature throughout the twentieth century, in addition to attempts to construct symmetric Venn diagrams for any number of terms (see [More, 1959] for instance). The last quarter of the twentieth century in particular saw a revival of interest in the geometry of Venn diagrams. In the second edition of his book (1983), Gardner added a long footnote to the chapter on "Logic Diagrams" in which he recorded some of the attempts to construct Venn diagrams for an arbitrary number of sets. As Venn had realized, starting with his 3-circle diagram, further sets can be added one at a time by threading the new set boundary through each of the existing subset areas.

Other methods are however possible and one of them, developed by the second author (A. W. F. Edwards) in the late 1980s, is of special interest here as it is related to both Venn and Carroll's diagrams. Although the author's early interest in the problem was mainly motivated by an event in celebration of Venn in his Cambridge college, Gonville and Caius, his explorations brought him in the end (though independently) to a Carrollian scheme of symmetry. The story of the emergence of the Edwards-Venn solution to the problem of depicting an arbitrary number of sets has been told in the book *Cogwheels of the Mind: The Story of Venn Diagrams* [Edwards, 2004]. In summary, the solution is a diagram in which it is obvious how the next set is to be added as a further "cogwheel" with twice the number of cogs of its predecessor, as may be seen when moving from 5 sets [Fig. 17] to 6 [Fig. 18]. The symmetry is Carroll-like, though in fact the diagrams were originally developed by contemplating the surface of a sphere as the universe to be partitioned. The important distinction is that each subsequent set is woven around the circle of the first set rather than the boundary of the immediately preceding set, so that it rides on the back of a smooth curve rather than one that is already cogged, thus minimizing the "comb-like" effect (to use Venn's expression already quoted). Had Carroll drawn his fifth set in [Fig. 14] more smoothly he might have felt encouraged to ponder further about the geometry.

Edwards-Venn diagrams have a particular structure that makes them different from their predecessors. Grünbaum's 5-set diagram alluded to earlier, for instance, possesses five-fold rotational symmetry, and similar

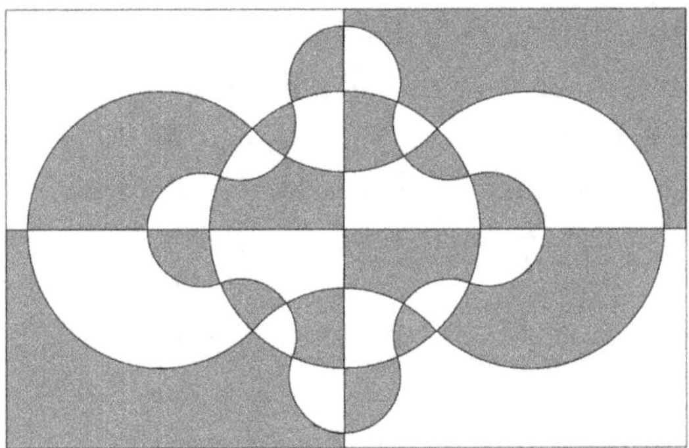

Fig. 17

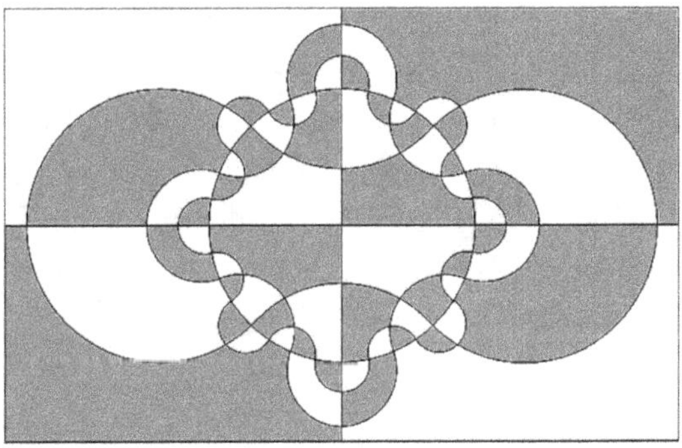

Fig. 18

diagrams have been found for 7-sets though not yet for 11-sets. It is known that only prime-number sets can yield such symmetrical solutions. All the developments so far described involve *simple* Venn diagrams, that is, ones in which set boundaries always intersect just two at a time. But from a logical point of view there is no reason for this restriction, and by rotating the sets of an Edwards-Venn diagram so that three or more boundaries intersect in the same point further diagrams may be formed (see Edwards [2004]). One such diagram has a curious pre-history, to which we now turn.

A GARDNER CURIOSITY

What was not realized by Venn or Jevons or Macfarlane, or indeed Gardner or anyone else until quite recently, was that by simply drawing lines round the letters of a Jevons logical alphabet or a Macfarlane logical spectrum so as to separate "absence" and "presence," a Venn diagram appears. This is most easily seen in the case of three classes, but the procedure is quite general [Fig. 19]. Though not a *simple* Venn diagram by virtue of the class boundaries intersecting more than two at a time, if the whole is encased in a circular universe it correctly displays the 8 regions formed by 3 intersecting sets. Moreover, it is entirely obvious how subsequent classes are to be added, one at a time, weaving around the imaginary diameter (which could itself be the boundary for another set).

It seems as though Gardner himself was not particularly interested in the problem of drawing diagrams for many classes, though as we have noted he was careful to record references to the progress that had been made by the time of his second edition. Remarkably, he was not aware that in 1971 he had published a diagram [Fig. 20] in another connection that was in essence the same as the Venn diagram of [Fig. 19], thus solving the problem himself! Nor did anyone else notice the class-diagram interpretation until after the same figure had been independently obtained, when the second author (A. W. F. Edwards) drew Gardner's attention to it in 1989. He replied:

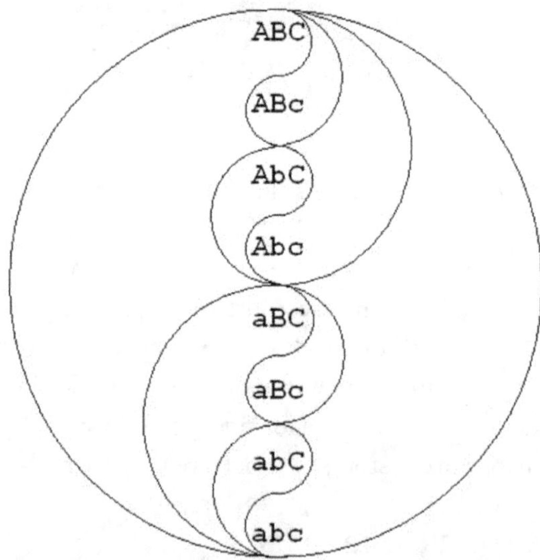

Fig. 19

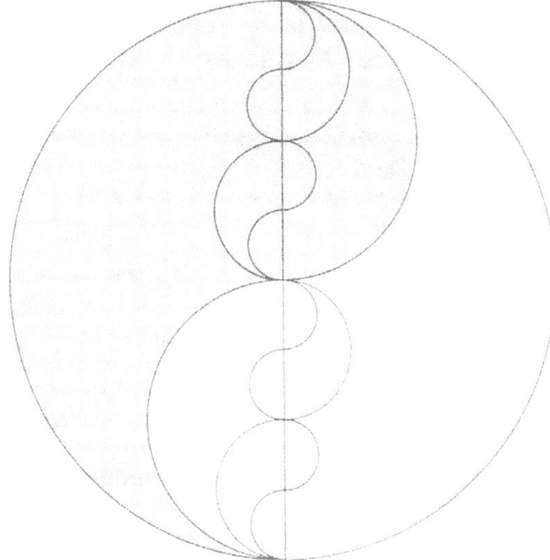

Fig. 20

> Thank you for calling my attention to the way that yin-yang pattern expresses a binary code. It certainly hadn't occurred to me, though it's obvious enough when it is pointed out. But that it is also a Venn diagram is much less obvious. The column on geometrical fallacies is reprinted in my Wheels, Life and Other Mathematical Amusements. Maybe I'll get a chance to add something on the pattern in a later edition of the book. [Gardner 1989]

In fact Gardner's 1971 diagram is for four sets, the fourth's boundary being the diameter rather than an equivalent wavy line.

Bibliography

Abeles, Francine F. (2007). "Lewis Carroll's Visual Logic." *History and Philosophy of Logic* 28, no. 1, February: 1–17.

Allwein, Gerard, and Jon Barwise, eds. (1996). *Logical Reasoning with Diagrams*. New York-Oxford: Oxford University Press.

Carroll, Lewis (1886). *The Game of Logic*. London: Macmillan (2nd ed., 1887).

_______ (1896). *Symbolic Logic. Part I: Elementary*. London: Macmillan (4th ed., 1897).

Edwards, A. W. F. (1989). "Venn Diagrams for Many Sets." *New Scientist* 121, no. 1646, 7 January: 51–56.

_______ (2004). *Cogwheels of the Mind: The Story of Venn Diagrams*. Maryland: Johns Hopkins University Press.

Euler, Leonhard (1768). *Lettres à une Princesse d'Allemagne*, vol. 2. Saint Petersburg: Imprimerie de l'Académie Impériale des Sciences.

Gardner, Martin (1958). *Logic Machines and Diagrams*. New York: McGraw-Hill (2nd ed., 1983).

_______ (1967). "Logic Diagrams," in P. Edwards (ed.), *Encyclopedia of Philosophy*, vol. 5. New York: Macmillan: 77–81.

_______ (1971). "Mathematical Recreations." *Scientific American,* April.

_______ (1989). Letter to A. W. F. Edwards, May 3 (private collection).

Grünbaum, Branko (1975). "Venn Diagrams and Independent Families of Sets." *Mathematics Magazine* 48, no. 1, January: 12–23.

Hamburger, Peter and Raymond E. Pippert (2000). "Venn Said It Couldn't Be Done." *Mathematics Magazine* 73, no. 2, April: 105–110.

Macfarlane, Alexander (1885). "The Logical Spectrum." *Philosophical Magazine* 19: 286–290.

Macula, A. J. (1995). "Lewis Carroll and the Enumeration of Minimal Covers." *The Mathematics Magazine* 68, no. 4, October: 269–274.

Marquand, Allan (1881). "Logical Diagrams for n Terms." *Philosophical Magazine* 12: 266–270.

Moktefi, Amirouche (2008). "Lewis Carroll's Logic," in Dov M. Gabbay and John Woods (eds.), *The Handbook of the History of Logic*, vol. 4: *British Logic in the Nineteenth Century*. Amsterdam: North-Holland: 457–505.

More, Jr., Trenchard (1959). "On the Construction of Venn Diagrams." *Journal of Symbolic Logic* 24, no. 4, December: 303–304.

Shin, Sun-Joo (1994). *The Logical Status of Diagrams*. New York: Cambridge University Press.

Venn, John (1880). "On the Diagrammatic and Mechanical Representation of Propositions and Reasonings." *Philosophical Magazine* 10: 1–18.

_______ (1881). *Symbolic Logic*. London: Macmillan (2nd ed., 1894).

Wakeling, Edward, ed. (2004). *Lewis Carroll's Diaries*, vol. 8. Clifford, Herefordshire: The Lewis Carroll Society (UK).

Probability Paradoxes

EUGENE SENETA

My interest in probability dates back to the year 1962, when I was a mathematics undergraduate at the University of Adelaide, South Australia. In that year I bought Martin Gardner's *Mathematical Puzzles and Diversions from Scientific American,* in a British edition of 1961. Its Section 5 had the same title as my present contribution.

Starting in 1965, as a young academic at the Australian National University, Canberra, in my teaching of statistics to Continuing Education classes, I frequently used to enliven the classes with the "Birthdays Paradox," which I had discovered from that Section 5, that the chance of two birthdays coinciding in a group of 24 or more is at least ½. And after, from reading the same Section 5 on Hempel's (also known as the Raven) Paradox, I saw that it could be used in the discussion of statistical inference within a geophysical

Eugene Seneta was born in March, 1941, in western Ukraine. He arrived in Australia in 1949, settling in Adelaide. In 1968 the Australian National University in Canberra conferred on him his Ph.D. in Statistics. He became Professor and Head of the Department of Mathematical Statistics at the University of Sydney in 1979, at which time he developed an interest in Lewis Carroll's probability problems. In 2007 Seneta was awarded the Hannan Medal in Statistical Science by the Australian Academy of Sciences.

problem on which Bob Anderssen and I were working.* Following up on the related philosophical literature listed in Gardner's suggestions for further reading, we eventually published a paper which cites, inter alia, some of this reading and our understanding of the paradox (Anderssen and Seneta, 1972). We spent many happy hours in the company of our young families on the counterintuitive logic that the simple hypothesis "All crows are black" could be confirmed by observation of almost anything, including our favorite: Western Australian black swans, since it was widely believed that all swans are white, observation of which would also confirm the hypothesis.†

In my first year at Sydney University I gave a course to high school teachers on elementary statistics, which requires the use of probability as the basis of statistical inference. After a class on the concepts of total probability and inverse probability, one of the listeners approached me with a problem on inverse probability from Lewis Carroll's *Pillow Problems.* A friend, Irene Buschtedt, then gave me as a gift Jean Gattegno's *Lewis Carroll: Fragments of a Looking Glass*. And so with the help and encouragement of my wife Ludmilla, of Irene, and later of the notable Carrollians Fran Abeles, the late Peter Heath, and, more recently, Amirouche Moktefi, I began my continuing study and interest in Lewis Carroll in general, and his probability work in particular (Seneta, 1993). Peter, a distinguished philosopher with a special interest in Kant and author of *A Philosopher's Alice*, and I spent meals in the University of Virginia environs over Carrollian topics, and together attended my only meeting (at Winston-Salem, N.C.) of the Lewis Carroll Society of North America. To my regret, Martin Gardner, author of *The Annotated Alice*, could not attend.

Martin Gardner does have in one of his books a section entitled *The Games and Puzzles of Lewis Carroll*, but this does not appear to involve any of Dodgson's probability problems. There is, however, a connection between one of Dodgson's probability problems (on the broken rod) and Martin Gardner's (1966) account of two problems in his book *More Mathematical*

* In simplest terms, the statement "All ravens are black" is logically equivalent to "Everything that is not black is not a raven." Hence (paradoxically) *anything* you observe that is not black and not a raven can be considered "proof" of the statement.

† On hearing of Martin Gardner's death, my son Bodie emailed me from Cambridge that he had used the same book after we moved to Sydney in 1980, spending much happy time constructing hexaflexagons, the topic of its Section 1.

Puzzles and Diversions within Section 19: *Probability and Ambiguity*. I turn to this context to now create variations on a theme, in memory of Martin Gardner.

LEWIS CARROLL'S BROKEN RODS

Lewis Carroll's Pillow Problem No. 45 reads as follows:

> If an infinite number of rods be broken: find the chance that one at least be broken in the middle.

His solution, "0.6321207 . . . ," begins by dividing mentally each rod into (n+1) equal parts, where n is an odd number, so there is a middle point, and assuming the n points of division are the only points where the rod can break, each breakpoint is equally likely. The probability that a rod will break in the middle is $1/n$, so that the probability of the complementary event, that it will break elsewhere is $(1 - 1/n)$. Now, taking the number of rods to be the same (!) as the number of breakpoints, the probability that all rods will break at a point *not* the middle is $(1 - 1/n)^n$ so the probability of at least one breaking in the middle is $1 - (1 - 1/n)^n$, and as n increases to infinity, this approaches $1 - e^{-1} = 0.6321207\ldots$

This is actually the correct answer to another problem: If a single rod can "break" into a number of pieces from 1 to n+1, with any actual breaks occurring at the n specified breakpoints only, what is the probability that there is at least one break? If we call a break at the ith breakpoint a *success*, the probability of at least one success is $1 - (1 - 1/n)^n$, which in the limit is $1 - e^{-1}$. This is the method of obtaining by discrete approximation the probability of at least one "event" occurring in a time interval of total length 1 in a purely random process over time. In such a process in continuous time the probability of an event occurring in an interval of small length δt is δt. (More generally it is taken as proportional to δt, to give the general form in the so-called Poisson process.)

In a probability problem where outcomes of a random experiment occur as points on a sample space of such points, such as a finite line interval, or an area such as a triangle, if the probability of a subset of such points (a subinterval, or area within the triangle) is exactly equal to the length or area of a subset, the probability is said to have a uniform distribution over the sample space. Such problems were generally approached by subdividing the sample space into a finite collection of equally probable outcomes, say n. In such a situation counting techniques (combinatorial methods) could be applied, together with the classical definition of probability as the proportion of outcomes favorable to an event whose probability is sought, and/or the simple rules of probability such as the addition rule, the complementarity event rule; and the multiplication rule (for independent events). The answer to the continuous problem was then obtained by letting n approach infinity.

Martin Gardner would have enjoyed the ambiguity with which Dodgson's problem No. 45 could be endowed, since in his article *Probability and Ambiguity* he focused on two problems to do with broken rods from the book of problems *Choice and Chance* by William A. Whitworth (1840–1905), and an accompanying book by the same author of hints and solutions (Whitworth, 1897).

Whitworth's *Choice and Chance* was, as now seems even more likely, one of the two books on probability in the reign of Queen Victoria from which Dodgson learned about probability manipulations.

WHITWORTHS'S BROKEN RODS

The two specific problems (among many on the broken rod) are expressed by Whitworth as a single two-part one:

> 677. If a straight line be broken at random into three parts the odds are 3 to 1 against their making the sides of a triangle; but if the line be first broken into two parts and then the larger portion be broken into two parts, the odds are 2 to 1.

Martin Gardner (1966) in his Section 19, actually paraphrases Whitworth's (1897), pp. 202–203, solution to both these parts, which is as follows:

> 677. Make an equilateral triangle ABC of altitude equal to the given straight line. Bisect the sides in A′, B′, C′. The trilinear coordinates of any point P within the triangle will represent segments into which the line may be divided. (See at the end of the volume the note on Qn. 632.) If no two are to be greater than the third, P must be within Δ A′B′C′. This area is one-fourth of the whole. [Therefore] Chance = $^1/_4$. Odds are 3 to 1 against.

> (ii) Let the perpendicular from P to BC represent the shorterof the two parts into which the line is first divided. P musttherefore lie on the area BC′B′C. But the favorable area is A′B′C′ as before. [Therefore] Chance = $^1/_3$. Odds are 2 to 1 against.

Both his solutions depend on area arguments using the supposed uniform probability distribution over the points of (differing) sample spaces. But this uniformity does not hold for part (ii), even though the triangular "event" whose probability is sought is the same in both parts. For part (ii) the ratio justification for probability is thus not appropriate, and the answer, $^1/_3$, is wrong.

Gardner (1966) reflects on this error in his Addendum on pp. 176–178, where he cites Whitworth (1897) explicitly as the source of his own error in solution, and cites numerous correct solutions from readers, including one from a textbook by Sokolnikoff and Redheffer (1958), p. 636. He gives an honest and salutary warning on his p. 176:

> In giving the second version of the broken stick problem I could not have picked a better illustration of the ease with which experts blunder on probability computations, and the dangers of relying on geometrical diagrams.

Yet such geometrical approaches, as in the solution of the first part, under propitious conditions such as uniformity and symmetry, often embody an elegance in their "surprise attack in mathematical problems" (this phrase is actually the title of L. A. Graham's book of 1968, in which the solution to the first part is discussed again on pp. 98–99). It is such elegance which appeals to puzzle constructors, solvers, and lateral thinkers such as Martin Gardner and his many kindred spirits. The second problem, however, cannot be completely resolved without the mathematical tool of elementary integral calculus.

Whitworth (1897), pp. 232–233, in his "note on Qn. 632" was at pains to make plausible, using a discrete approximation argument,

> that the points representing an equable distribution of fractures are equably distributed over the area. This is easily seen when we consider a magnitude capable of only a finite number of

> fractures, e.g., we may consider a row of 8 elements to be divided at random into three parts. If we take all possible fractures, viz. 8, 0, 0; 7, 1, 0; 6, 1, 1; etc. and mark the corresponding points on the triangle, we find that we mark once and once only the angular points of 256 small triangles, and these are all equal, and fill up the space of the original triangle.

There may be a small error in Whitworth's calculation of the number of small triangles, and this would have been noticed, though not mentioned, by Martin Gardner. There appear to be 45 points in all of the form (x, y, z) where x, y, z are non-negative integers and $x + y + z = 8$, and the corresponding number of triangles is $64 = 2^6$, whereas $256 = 2^8$. If one takes just 4 points rather than 8, the number of small triangles is $16 = 2^4$, which is $(2^2)^2$. Whitworth may have extrapolated to $2^8 = (2^4)^2$ instead of to $(2^2)^3$, or changed the problem without changing the answer.

HISTORY AND ANALYTICAL APPROACH

The problem of the broken rod in its variants and generalizations is an old one. If its history is thought to begin with Whitworth's *Choice and Chance* in one of its editions (the first appeared in 1867, the second in 1870), there is a rich and presumably "independent" French-language literature (Seneta and Jongmans, 2005). Whitworth's first part of problem 677 was proposed and solved by Lemoine (1872), who considered the rod divided conceptually into 2m equal segments and, using combinatorial counting and then taking the limit as $m \to \infty$, found the correct probability: $1/4$. Immediately following him, Halphen (1872) solved a more general problem in a particularly elegant way using simple rules for probability manipulation.

We reproduce Halphen's solution for the triangle problem: it would surely have delighted Martin Gardner, and so perhaps please the readers of this bouquet in his honor:

The fulcrum on which the argument turns is that the lengths of 3 pieces can form a triangle if and only if each length is less than ½ the total. (We leave it to the reader to verify this.)

To formalize the argument, let the points of breakage be X_1, X_2 each being an independently and uniformly distributed point chosen along the length of the stick, which is taken to be 1. Measuring from the left end of the stick (taken as origin) let the positions be $Y_{(1)}$ and $Y_{(2)}$, so that $0 < Y_{(1)} < Y_{(2)} < 1$, where $Y_{(1)} = \min(X_1, X_2)$, and $Y_{(2)} = \max(X_1, X_2)$. Put now $W_1 = Y_{(1)}$, $W_2 = Y_{(2)} - Y_{(1)}$, $W_3 = 1 - Y_{(2)}$.

The required probability is then:

$$P(W_1 < ½, W_2 < ½, W_3 < ½)$$
$$= 1 - P(W_1 > ½ \text{ or } W_2 > ½ \text{ or } W_3 > ½)$$
$$= 1 - (P(W_1 > ½) + P(W_2 > ½) + P(W_3 > ½))$$

using in turn the complementation rule; and the addition rule for probabilities since the events being considered are mutually exclusive;

$$= 1 - 3P(W_1 > ½)$$

since the three probabilities are the same (the "W_i"s are identically distributed);

$$= 1 - 3P(Y_{(1)} > ½) = 1 - 3P(X_1 > ½, X_2 > ½)$$
$$= 1 - 3P(X_1 > ½)P(X_2 > ½)$$

by the multiplication rule;

$$= 1 - ¾ = ¼$$

since X_1, X_2 are uniformly distributed on $(0,1)$. It is clear how this argument would generalize to n breakage points, to form a convex polygon with $(n+1)$ sides.

Finally, we offer another analytical approach to the first problem. The expression of the event obtained analytically before the evaluation of probability is exactly parallel in the first and second problems, as we shall show below, and as Whitworth and Martin Gardner recognized geometrically. However, it is now more clearly seen at what point the procedures differ, and how to evaluate the probability, although there is nothing new in our evaluation. We shall denote by the symbol Δ the event that a triangle can be formed. The meaning of X_1 and X_2 is as before. And we use Halphen's device that a triangle can be formed if and only if each piece has length < ½. Then:

$$P(\Delta) = P(\Delta, X_1 > ½) + P(\Delta, X_1 < ½)$$
$$= P(\Delta, X_2 > X_1, X_1 > ½) + P(\Delta, X_2 < X_1, X_1 > ½)$$
$$+ P(\Delta, X_2 > X_1, X_1 < ½) + P(\Delta, X_2 < X_1, X_1 < ½)$$
$$= 0 + P(\Delta, X_2 < X_1, X_1 > ½) + P(\Delta, X_2 > X_1, X_1 < ½) + 0$$
$$= P(X_2 < X_1, X_1 > ½, X_2 < ½, X_1 - X_2 < ½, 1 - X_1 < ½)$$
$$+ P(X_2 > X_1, X_1 < ½, X_1 < ½, X_2 - X_1 < ½, 1 - X_2 < ½)$$
$$= P(½ < X_1 < X_2 + ½, X_2 < ½,) + P(½ < X_2 < X_1 + ½, X_1 < ½)$$
$$= 2P(½ < X_1 < X_2 + ½, X_2 < ½) \quad (1)$$

by symmetry.

For part (ii) let W be the distance from the left hand end to the break point of the longer piece. Then

$$P(\Delta) = P(\Delta, X_1 > ½) + P(\Delta, X_1 < ½)$$
$$= P(W < ½, X_1 - W < ½, 1 - X_1 < ½)$$
$$+ P(W < ½, 1 - X_1 - W < ½, X_1 < ½)$$
$$= P(X_1 - ½ < W < ½, X_1 > ½)$$
$$+ P(1 - X_1 - ½ < W < ½, 1 - X_1 > ½)$$
$$= 2P(½ < X_1 < W + ½, W < ½) \qquad (2)$$

by symmetry.

We can see from (1) and (2) that the event whose probability is to be found, as an area in two dimensional space, is the same. However, in (1) X_1, X_2 are independently distributed over (0,1), and (1) evaluates to ¼ directly by area comparison. In (2), given X_1 as a value greater than ½, the distribution of W is uniform over $(0, X_1)$, and the (conditional) probability of $X_1 - ½ < W < ½$ is therefore $(½ - (X_1 - ½)) / X_1$. This to be averaged over $X_1 > ½$ using the uniform distribution over (0,1) of X_1, and the result multiplied by 2 to evaluate (2):

$$2\int_{(½, 1)} (1 - x) / x \, dx = 2(\ln 2 - ½) = 0.38629436\ldots$$

References

Anderssen, R. S. and E. Seneta (1972). A simple statistical estimation procedure for Monte Carlo inversion in geophysics. II: Efficiency and Hempel's paradox. *Pure and Applied Geophysics* 96, no. 1: 5–14.

Gardner, M. (1961). *Mathematical Puzzles and Diversions from Scientific American*. London: G. Bell and Sons. [First published in 1959.]

Gardner, M. (1966). *More Mathematical Puzzles and Diversions*. Harmondsworth: Penguin.

Halphen, G. (1872). Sur un problème de probabilités. *Bulletin de Société Mathématique de France* 1: 256–258, 281.

Lemoine, E. (1872). Sur un question de probabilités. *Bulletin de Société Mathématique de France* 1: 39–40.

Seneta, E. (1993). Lewis Carroll's "Pillow Problems": On the 1993 centenary. *Statistical Science* 8, no. 2: 180–186.

Seneta, E. and Jongmans, F. (2005). The problem of the broken rod and Ernesto Cesàro's early work in probability. *Mathematical Scientist* 30, no. 2: 67–76.

Sokolnikoff, I.S. and Redheffer, R. M. (1958). *Mathematics of Physics and Modern Engineering*. New York: McGraw-Hill.

Whitworth, W. A. (1965). *Choice and Chance with One Thousand Exercises*. New York: Hafner. [Reprint of 5th edition of 1901.]

Whitworth, W. A. (1897). *DCC Exercises, Including Hints for Solution of All the Questions in Choice and Chance*. Cambridge: Deighton Bell.

Four Puzzles for Martin

RAYMOND SMULLYAN

I first knew Martin when we were students at the University of Chicago. He has since been a most wonderful and constant friend, and has helped me enormously in my career as a puzzle writer, once paying me the supreme compliment of telling me, "Your puzzles have charm."

Martin and I had many interests in common: not only puzzles, but magic, literature, and religion. I have been a professional magician and he was an excellent magician whose technique was far better than mine. As to literature, for example, both of us were fans of Lewis Carroll: he wrote *The Annotated Alice*, I wrote *Alice in Puzzle-Land*.[1]

Although our views on religion were very different, I have always been extremely fascinated by his religious writings. *The Flight of Peter Fromm* is the best religious novel I have ever read; it shows remarkable psychological insight. Less impressive, in my opinion, are the religious chapters of *The Whys of a Philosophical Scrivener*. Some of them struck me as too moralistic, and I had objections to several of his points, all of which I wrote to him about. For one thing Martin identified belief in God with belief in an afterlife, and I

Raymond Smullyan is an American mathematician, musician, magician, videographer, and the author of twenty-six books, including eleven books of recreational logic puzzles, such as What is the Name of This Book?, *which Martin Gardner pronounced "the most original, most profound, and most humorous collection of recreational logic and math problems ever written." His videos, piano playing, jokes, puzzles, and other activities can all be found at RaymondSmullyan.com. He is the subject of the 2001 documentary* This Film Needs No Title.

believe the two should be sharply separated. These days belief in an afterlife has declined far more than belief in God. I know many people who firmly believe in God and just as firmly disbelieve in an afterlife. I also know some who believe in an afterlife and who do not believe in God. Another point of disagreement is when he wrote, "Is there any significant difference between not believing in God and believing there is no God, or not believing in an afterlife and believing there is no afterlife?" I pointed out to Martin that there is indeed an enormous difference! One who doesn't believe in an afterlife but also doesn't disbelieve can at least have hopes that there may be an afterlife, whereas one who disbelieves can have no such hopes. Likewise with non-belief and disbelief in God.

I wrote these thoughts to Martin along with many others about his book and he graciously wrote me back a lovely letter which began, "Dear Raymond, I've now had a chance to read your critique carefully and set down reactions. I can't imagine anyone providing a more sympathetic, insightful, stimulating, or wiser critique. You have persuaded me I was wrong to blur the distinction between atheism and agnosticism. You are certainly right in saying that an agnostic can at least have hope, whereas a convinced materialist can't, and that is a major difference." Later on he wrote, "You have also convinced me that I linked too closely the belief in theism and an afterlife." He ended the letter with the following sweet paragraph: "As always, I love your humorous anecdotes and your informal writing style. I *very* much appreciate your taking my views so seriously. Not many do. At least there has been not one review of my book in any philosophical or theological journal, and only trivial reviews in a few newspapers and magazines."

As many of you know, for a long time there has been a "Gathering for Gardner" in Atlanta every other year, and I have attended almost all of them. At one of them I had intended to play a trick on Martin, but when I came I was disappointed to find that he was unable to attend. I then played the trick on another attendee with a slightly different ending. I handed him two ten-dollar bills and said, "I am about to make a statement. If the statement is false, then you must promise me to give me back one of the bills and keep the other. But if the statement is true, then you must keep them both. Agreed?" He agreed. I then made a statement such that the only way he could keep to the agreement is by owing me a thousand dollars! Can you guess what statement would work? [Puzzle #1. Answers are given at the end of this article.]

I then said, "I feel a little guilty about this, and so I want you to have a chance to win your money back again, and so I will give you back your thousand dollars, providing you agree to truthfully answer me a yes/no question. Do you agree?" He agreed I then asked him a question such that the only way he could keep his word was by owing me a million dollars instead of a thousand. What question would work? [Puzzle # 2]

I then said, "I really feel ashamed of myself for doing this, and so I will now give you a better chance. Again I will ask you a yes/no question, but this time you don't even have to answer truthfully! Your answer can be either true or false as you choose! There is no possible way I can con you now, right?" Well, there is a way I could—and did—con him. My question was such that regardless of whether he would answer truthfully or falsely. he would now owe me a billion dollars instead of a million! What question would work? [Puzzle #3]

Next, I used a trick I learned from one of Martin's books. I said to the victim, "I will give you a fifty percent chance of winning back your billion dollars, but for that I charge a nickel extra. Isn't it worth a nickel for a fifty percent chance of getting back your billion dollars? (This of course got a good general laugh.) I then wrote something on a piece of paper and folded it and gave it to someone to hold and I told the victim that I had just written down a description of an event which will or will not take place in this room within the next fifteen minutes. You are to guess whether or not the event will take place. If you guess correctly, you will get back your billion dollars, but if you guess wrong, you won't. In either case you still owe me a nickel. Now, your chance of guessing correctly is obviously fifty percent, correct? He agreed. I then told him to decide in his mind whether or not the event will take place, and I handed him a pen and a piece of paper and said, "I want a written record of this and so if you believe the event will take place, write down YES, and if you believe it won't, write down NO. I will turn my back and won't see what you write." After while I turned around again and asked him if he had written his answer. When he told me that he had, I then said, "Then you've lost!" What I wrote was something such that regardless of what he would answer, he would lose! Can you guess what I wrote? [Puzzle #4]

I have pulled that entire routine on several occasions and each time I used a different method of getting the victim to get his billion dollars back. Now, if Martin had been present at the gathering I would have omitted the last trick, which he already knew, and when he owed me a billion dollars, I would have said:" I'll tell you what, Martin, I will trade you the whole billion dollars for one kiss from Charlotte." (Charlotte was his lovely wife.) After the gathering was over, I phoned Martin and told him what I had planned to do, and we both had a good laugh over it. I told him to please tell Charlotte of my plan, which he did and later told me that she was quite amused.

Martin was the one who introduced me to the Newcomb Paradox. For those who don't know it, I present it in the following form: You are in front of a chest with two drawers, in each of which there is either a hundred dollar bill or a thousand dollar bill. Both drawers have the same amount, and so there is either two hundred dollars or two thousand dollars all told. You are given the option of taking the money from just the bottom drawer or the money from both drawers. Which would you choose? Of course you would choose both drawers. Now, is there anything I could tell you that would get you to change your mind? Most people answer NO.

Well, what I didn't tell you is that there is an omniscient being—a god or a computer—who knows the future and how you will choose and has complete control of how much money goes into the drawers. If it (he? she?) predicts that you will choose both drawers, then it will put a hundred dollars in each drawer, but if it predicts that you will choose just the bottom drawer, then it will put a thousand dollars in each. Would that additional information change your mind? There are two schools of thought about this.

One school says that under these circumstances you should choose just the bottom drawer, thus getting a thousand dollars, whereas if you choose both drawers you will get only two hundred. The other school says *nonsense!* the money is already there, and there is twice as much in both drawers than in only one drawer, and so you should of course choose both drawers. That is the paradox. When Martin told me of this, I said that I would of course choose just the bottom drawer, thus getting a thousand instead of two hundred. He replied, "Don't be too hasty, Raymond! Imagine that the backs of the drawers were made of glass and that friends of yours were standing behind and could see how much the money was in them. They were rooting for you and wanted you to get as much money as possible. Wouldn't they hope that you would choose both drawers?" I didn't know how to answer at the time, but some time later realized what my answer should be, which is that if my friends knew about the predictor

then they would know how I would choose, and could therefore not hope that I would choose otherwise.

Martin discusses this paradox in one of his books,[2] and came to the conclusion that the paradox proves that there cannot be a perfect predictor. With all due respects to Martin, I must disagree, and so I came up with a variant of the paradox in which the predictor doesn't even appear!

As before, there is either a hundred or a thousand dollars in each drawer. Now consider the following proposition:

> Either you will choose both drawers and there is a hundred dollars in each, or you will choose just the bottom drawer and there is a thousand dollars in each.

Please note that nothing is mentioned about a perfect predictor. The function of the predictor was to give plausibility to that proposition. If the proposition is inconsistent, that would support Martin's contention. Is the proposition consistent? First I will prove that it is inconsistent, and then I will prove that it is consistent. That's pretty paradoxical, isn't it?

To prove that the proposition is inconsistent, we see that on the one hand you will get more money if you choose just the bottom drawer—one thousand versus two hundred—but on the other hand you will get more by choosing both drawers, since there is twice as much money there than in just the bottom drawer. This is a clear inconsistency.

Now, to prove that the proposition is consistent, it suffices to show that there could be a circumstance in which it is true. Well, it is certainly possible that you choose just the bottom drawer and that there is a thousand dollars in each drawer (alternatively, it is possible that you choose both drawers and that there is a hundred in each) and so the proposition is consistent after all! That is my version of the paradox in which the predictor doesn't appear.

In conclusion, let me tell you two delightful incidents: As many of you know, Martin was great on playing April Fools jokes. Well, I once pulled one on him, which he fell for—for a while. I phoned him one April 1 and said, "What do you think of that remarkable article in today's *New York Times* about Leonardo da Vinci?" When he told me that he hadn't read it, I said: "There is now absolute evidence that da Vinci was really a woman! Her real name was Maria da Vinci." Martin was at first amazed, then expressed skepticism as to whether the article should be trusted, and then suddenly realized that it was April 1.

On another occasion I phoned him about his book *Confessions of a Psychic*, which he wrote under the pseudonym Uriah Fuller, in which he so cleverly exposed the fraudulence of fake psychics. In a terrible voice I said, "LOOK! THIS IS URIAH FULLER AND I WANT YOU TO KEEP OUT OF MY TERRITORY, SEE!" In his sweet, gentle voice he said, " Oh, hi, Raymond."

SOLUTIONS

1. I said, "You will give me either one of the two bills or a thousand dollars." If the statement were false, he would have to give me one of the two bills as agreed, but doing so would make the statement true, which is a contradiction: hence the statement can't be false; it must be true. Thus it is true that he must give me either one of the bills or a thousand dollars. But the agreement was that if the statement is true, he must keep both bills, and so his only option is to give me a thousand dollars!

2. I asked him whether he would either answer NO to the question, or pay me a million dollars. Thus I am asking whether one of the following

alternatives holds: (1) he will answer NO; (2) he will pay me a million dollars. If he answers NO he is denying that either alternative holds when in fact (1) did hold: Hence he would be wrong, and so the only truthful answer is YES, which means that he affirms that one of the two alternatives holds, but now it cannot be (1), hence it must be (2), which means that he now owes me a million dollars!

3. I said that he could answer either truthfully or falsely; I never said that he could answer paradoxically! I asked him, "Is YES the correct answer to this question if and only if you pay me a billion dollars?" If he doesn't pay me a billion dollars, my question reduces to "Is NO the correct answer to this question?" and neither YES nor NO could be either true or false without contradiction!

4. I wrote, "You will answer NO." If he answers NO then he is denying that the event will take place, when in fact it did. If he answers YES then he is affirming that the event would take place, when in fact it didn't. Thus he loses either way.

   Of all the tricks I learned from Martin Gardner, this is one of my very favorites!

1 *Alice in Puzzle-Land: A Carrollian Tale for Children Under Eighty* (William Morrow & Co., 1982)

2 *Knotted Doughnuts and Other Mathematical Entertainments* (W. H. Freeman, 1986).

Charles Dodgson's Geometry

ROBIN WILSON

Throughout his life Charles Dodgson was fascinated by all things geometrical. In *A New Theory of Parallels* (1888), he waxed lyrical about Pythagoras's theorem on right-angled triangles:

> It is as dazzlingly beautiful now as it was in the day when Pythagoras first discovered it, and celebrated its advent, it is said, by sacrificing a hecatomb of oxen [a hundred]—a method of doing honour to Science that has always seemed to me slightly exaggerated and uncalled-for. One can imagine oneself, even in these degenerate days, marking the epoch of some brilliant scientific discovery by inviting a convivial friend or two, to join one in a beefsteak and a bottle of wine. But a hecatomb of oxen! It would produce a quite inconvenient supply of beef.

Robin Wilson is an Emeritus Professor of Pure Mathematics at the Open University, in Milton Keynes, U.K.; Emeritus Professor of Geometry at Gresham College, London; and President-elect of the British Society for the History of Mathematics. He is involved with the popularization and communication of mathematics and its history, and has written and edited many books on those subjects, including Introduction to Graph Theory, Four Colours Suffice, *and* Lewis Carroll in Numberland. *He has received two awards from the Mathematical Association of America for his "outstanding expository writing."*

A year later, in *Sylvie and Bruno*, he perpetrated some dreadful geometrical word-play:

> PROFESSOR: (drawing a long line on the black board, and marking the letters "A", "B", at the two ends, and "C" in the middle) If AB were to be divided into two parts at C—
>
> BRUNO: It would be drownded.
>
> PROFESSOR: What would be drownded?
>
> BRUNO: Why the bumble-bee, of course! And the two bits would sink down in the sea!

And forty years earlier, on learning of six new additions to the Dodgson household, he wrote to his sister Elizabeth from Rugby School suggesting a geometrical name for one of them:

> I am glad to hear of the 6 rabbits. For the new name after some consideration I recommend Parellelopipedon—It is a nice easy one to remember, and the rabbit will soon learn it.

(A *parallelopipedon*, as he should have written it, is the Greek word for a parallelepiped, a skewed box bounded by parallelograms.)

But even before going to Rugby, the young Charles had demonstrated that he took his geometrical studies very seriously. A remarkable two-page document, written at the age of twelve (Figure 1), shows that he had already become conversant with the basic ideas of formal geometry, as exhibited in classical works such as Euclid's *Elements*, the most important geometry book of his (and all) time.

HERE'S LOOKING AT EUCLID

Throughout his life, Dodgson was involved with the thirteen books that comprise the *Elements*. As a schoolboy he had to learn propositions from them, and as a young lecturer he had to teach from them. Many of his publications were devoted to explaining them, championing them when they were under threat, or referring to them in a humorous way, as in *Hiawatha's Photographing*, where he described Hiawatha's box camera as looking "all squares and oblongs, like a complicated figure in the second book of Euclid." But why was the *Elements* so important to Dodgson?

Euclid lived around 300 BC in Alexandria (now in Egypt) and wrote the *Elements* there. For over 2000 years—from the ancient Greek academies

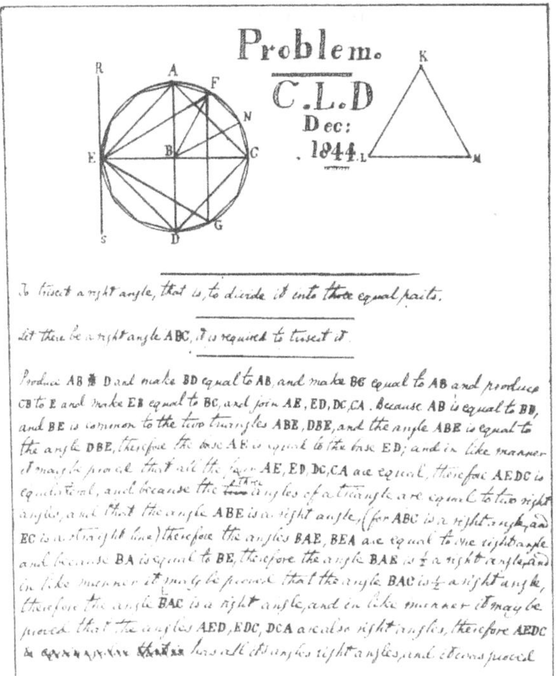

Fig. 1: How to trisect a right angle

and the universities of medieval Europe to the private schools of Victorian England—this text was used to teach geometry and train the mind. Indeed, the *Elements* is believed to be the most printed book of all time, after the Bible: during the nineteenth century over two hundred editions were published in England alone, with one popular version selling over half a million copies.

The many available editions of the *Elements* differed greatly in style and content. Over the Easter vacation of 1855 Charles Dodgson started teaching geometry to Louisa, the most mathematically gifted of his sisters:

> Went into Darlington—bought at Swale's, Chamber's Euclid for Louisa. I had to scratch out a good deal he had interpolated, (e.g. definitions of words of his own) and put some he had left out. An author has no right to mangle the original writer whom he employs: all additional matter should be carefully

> distinguished from the genuine text. N. B. Pott's [sic] Euclid is the only edition worth getting. . . .

(Robert Potts's edition, Dodgson's preferred choice, was *The School Edition, Euclid's Elements of Geometry, The First Six Books*.)

A STUDENT AT OXFORD

In May 1850, with his impressive Rugby School record, Charles Dodgson travelled to Oxford for his matriculation examinations in Latin, Greek, and mathematics. He was accepted at Christ Church, his father's former college, and went into residence on January 24, 1851.

The Oxford University course required students to sit three main examinations—*Responsions*, *Moderations*, and *Finals*. Those offering the geometry paper in Responsions would need to know the material of *Books I* and *II* of the *Elements*, while passmen (those who study for a degree without Honours) taking the Euclid paper in Moderations would also study *Book III*, and classmen taking Mathematics Finals would continue up to *Book VI*.

Responsions (or "Little-Go") was the first hurdle. Taking place twice a year, it included papers on Latin and Greek, a paper of grammatical questions, basic arithmetic (up to the extraction of square roots), and a paper on algebra or geometry. Most students attempted it after a year or more, but Dodgson was better prepared than most and took it in June 1851.

After a long summer vacation with his brothers and sisters, he returned to Oxford to begin preparations for Moderations, attending lectures on pure geometry by the Savilian Professor of Geometry (the Rev. Baden Powell) and studying with the College Lecturer (Robert Faussett). The examination took place in November 1852 and contained papers on the four Gospels, one Greek and one Latin author, and a paper on logic or geometry and algebra. In addition, Honours candidates took a paper in pure mathematics. Dodgson achieved a First Class in Mathematics and a Second Class in the Classics.

His Finals examinations were in two parts. The first, on the classical languages and literature, took place in the Easter Term of 1854, and he achieved only Third Class status. The second part, in December 1854, was in his chosen area of mathematics, for which the minimum requirement was "The first six books of Euclid, or the first part of Algebra." Candidates for

Honours, like Dodgson, were also required to study "Mixed as well as Pure Mathematics," involving a range of topics from the differential and integral calculus to astronomy and optics.

Dodgson performed well, coming top of the list of students achieving a First Class in Mathematics.

A YOUNG TEACHER

While studying at Christ Church, Charles Dodgson so impressed his tutors that he was awarded a Studentship. This entitled him to live in College for life, provided that he trained for the priesthood and remained unmarried. However, he now had to support himself by taking on private pupils:

> Got a note from Leighton, a gentleman commoner, who wishes to be taught some arithmetic for his Little-Go (which comes in about a fortnight)—as well as the second book of Euclid. We settled that he is to come an hour an evening, and began at once.

Two weeks later Baldwin Leighton passed his Responsions, and on receiving the £5 payment, Dodgson recorded that this was the first money he had ever earned.

By the Easter Term of 1855 the number of his private pupils had increased to fourteen. Robert Faussett was about to leave to fight in the Crimean War, and Dodgson considered that the experience his teaching gave him would increase his chances of getting the Mathematical Lectureship. He organized his students into tutorial groups, remarking ruefully that his fifteen hours of teaching each week would be "a remedy against idleness, such as I could never have devised for myself."

Dodgson did indeed gain the Mathematical Lectureship, which he then held for twenty-five years. He later recalled that

> I had the Honourmen for the last two terms of 1855, but was not full Lecturer till Hilary, 1856. . . . I gave my first Euclid Lecture in the Lecture-room on Monday January 28, 1856. It consisted of 12 men, of whom nine attended.

Opinions differ widely as to how good a teacher Dodgson was. One former undergraduate was complimentary:

> I was up at Christ Church as an undergraduate early in the eighties, he being my mathematical tutor and certainly his methods of explaining the elements of Euclid gave me the impression of being extremely lucid, so that the least intelligent of us could grasp at any rate 'the Pons Asinorum'.

(The Pons Asinorum, also known as the isosceles triangles theorem, is the name given to Euclid's fifth proposition in Book I of the *Elements*.) But another remembrance was more cryptic:

> . . . glancing at a problem in Euclid which I had written out, he placed his finger on an omission. "I deny you the right to assert that." I supplied what was wanting. "Why did you not say so before? What is a corollary?" Silence. "Do you ever play billiards?" "Sometimes." "If you attempted a cannon, missed, and holed your own and the red ball, what would you call it?" "A fluke." "Exactly. A corollary is a fluke in Euclid. Good morning."

DODGSON'S PAMPHLETS

Although Dodgson greatly admired Euclid's *Elements*, he recognized that it contained gaps, inaccuracies, and inconsistencies. To help his pupils overcome these, he produced a number of short mathematical pamphlets that clarified the text, suggested alternative approaches, and included exercises for his pupils to try. Over the years he became an enthusiastic producer of pamphlets, publishing more than two hundred of them on a wide range of topics.

His first mathematical pamphlet was *Notes on the First Two Books of Euclid* (1860), costing sixpence and designed to help those taking Responsions. These were subsequently expanded into book form, *Euclid, Books I, II*, which appeared in an unpublished private version in 1875 and were published in 1882, eventually running to eight editions. Their aim was

> to show what Euclid's method really is in itself, when stripped of all accidental verbiage and repetition. With this object, I have held myself free to alter and abridge the language wherever it

> seemed desirable, so long as I made no real change in his methods of proof, or in his logical sequence. The result is that the text of this Edition is (as I have ascertained by counting the words) less than five-sevenths of that contained in the ordinary Editions.

Associated with these *Notes* was a useful pamphlet entitled *The Enunciations of Euclid I, II* (1863), listing all the main definitions and propositions needed for Responsions; he later expanded this to *The Enunciations of Euclid I–VI* for those taking Finals.

In 1868 he produced a pamphlet for the Pass Schools, entitled *The Fifth Book of Euclid Treated Algebraically*, in which he explained each definition with examples and recast each proposition in a more accessible algebraic form. For simplicity he omitted the theory of incommensurable numbers as being inappropriate for undergraduates.

While these pamphlets and books were generally well received and widely used, not everyone was enthusiastic. In particular, Robert Potts, editor of Dodgson's preferred version of the *Elements*, had concerns about spoon-feeding the readers:

> I have had considerable experience in dealing with minds of low logical power, and have found that studies may be made so easy and mechanical as to render thought almost superfluous.

THE DYNAMICS OF A PARTI-CLE

Not all of Dodgson's geometrical endeavors were so serious. His witty, and often reproduced, pamphlet *The Dynamics of a Parti-cle* satirized the election for the Oxford University parliamentary seat in July 1865.

In the Victorian era Oxford University was represented by two Members of Parliament in London. In 1865 the candidates for these seats were the sitting candidate William Ewart Gladstone, who had represented the University for eighteen years but who was too liberal for Dodgson, the more conservative Gathorne Gathorne-Hardy who was Dodgson's preferred choice, and a third candidate, Sir William Heathcote.

Much of the pamphlet is a parody on Euclid. For example, Book I of the *Elements* opens with five postulates, of which three are:

> Let it be granted, that a line may be drawn from any point to any other point.

> That a finite line may be produced to any extent.
>
> That a circle may be drawn about any point, and at any distance from that point.

In the pamphlet, these appear as:

> Let it be granted, that a speaker may digress from any one point to any other point. That a finite argument (i.e. one finished and disposed of) may be produced to any extent in subsequent debates.
>
> That a controversy may be raised about any question, and at any distance from that question.

Carroll's treatise includes the observation:

> When this is effected, it will be found most convenient to project WEG to infinity.

This indeed happened: Heathcote and Gathorne-Hardy were duly elected.

EUCLID AND HIS MODERN RIVALS

In most English private schools, the Victorian curriculum consisted mainly of the classical languages, together with some divinity. For those that also taught mathematics, Euclid's geometry was the standard fare, being regarded as the ideal vehicle for teaching young men how to reason and think logically. Based on absolutes, the study of geometry fitted well with the classical curriculum, providing a suitable training for those expecting to go on to Oxford and Cambridge Universities and the Church. The *Elements* thus became an important constituent of examination syllabuses, being required also for entrance to the civil service and the army.

However, it was a time of change. A growing middle class demanded a more practical approach to mathematics, and the traditional classical education became increasingly irrelevant. In his 1869 Presidential address to the British Association for the Advancement of Science, the mathematician James Joseph Sylvester was forthright in condemning the old ways. Soon thereafter an Anti-Euclid Association was formed, and January 1871 saw the foundation of the Association for the Improvement of Geometrical Teaching, which took on the task of producing new geometry syllabuses and texts.

Charles Dodgson, an outspoken advocate for Euclid's *Elements*, was bitterly opposed to this new Association and its aims. In 1879 he wrote a remarkable work, *Euclid and His Modern Rivals*, in which he skillfully compared the *Elements*, favorably in each case, with thirteen well-known rival texts.

Attempting to reach a wider audience, Dodgson cast his book as a play in four acts:

> It is presented in dramatic form, partly because it seemed a better way of exhibiting in alternation the arguments on the two sides of the question; partly that I might feel myself at liberty to treat it in a rather lighter style than would have suited an essay, and thus to make it a little less tedious and a little more acceptable to unscientific readers.

There are four characters: Minos and Radamanthus (two of the three judges in Hades, appearing here as hassled Oxford examiners), Herr Niemand (the phantasm of a German professor who "has read all books, and is ready to defend any thesis, true or untrue"), and the ghost of Euclid himself.

The book was a *tour de force*, exhibiting Dodgson's intimate knowledge and deep understanding of Euclidean geometry. But ultimately it failed to achieve its aim. In 1888 Oxford and Cambridge, fearing that abandoning Euclid would reduce the examination systems to chaos, reluctantly agreed to accept proofs other than Euclid's, provided that they did not violate Euclid's ordering of the propositions. By 1903 this restriction too had disappeared, when they agreed to accept any systematic treatment. A formal approach to geometry continued to be taught in many high schools until about the 1960s, when it was quietly dropped. Today, few schoolchildren are aware of the Euclidean approach to geometry.

DODGSON'S HEXAGON

A major difficulty with the *Elements* was Euclid's fifth postulate (known to Victorians as his 12th axiom). The other four postulates are short and simple, but this one is more complicated:

> If a straight line falling on two straight lines makes the interior angles on the same side less than two right angles, the two

> straight lines, if produced indefinitely, meet on that side on which are the angles less than the two right angles.

This says that if angle A + angle B is less than 180° in the following picture, then the two lines must eventually meet, as illustrated.

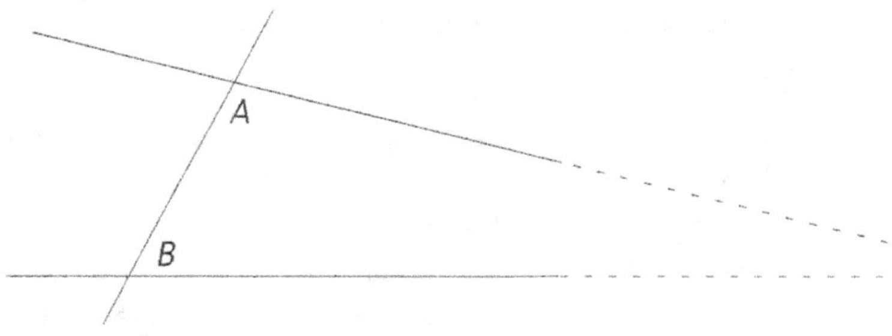

Fig. 2

In his amusing introduction to *The Dynamics of a Parti-cle*, Dodgson referred to the fifth postulate as "a striking illustration of the advantage of introducing the human element into the hitherto barren region of Mathematics" in the well-known paragraph:

> It was a lovely Autumn evening, and the glorious effects of chromatic aberration were beginning to show themselves in the atmosphere as the earth revolved away from the great western luminary, when two lines might have been observed wending their weary way across a plane superficies. The elder of the two had by long practice acquired the art, so painful to young and impulsive loci, of lying evenly between his extreme points; but the younger, in her girlish impetuosity, was ever longing to diverge and become a hyperbola or some such romantic and boundless curve. They had lived and loved: fate and the intervening superficies had hitherto kept them asunder, but this was no longer to be: *a line had intersected them, making the two interior angles together less than two right angles.* It was a moment never to be forgotten, and, as they journeyed on, a whisper thrilled along the superficies in isochronous waves of sound, "Yes! We shall at length meet if continually produced!"

For two thousand years, generations of mathematicians tried to deduce the fifth postulate from the others. In *Euclid and his Modern Rivals* Minos observes:

> MINOS: An absolute proof of it, from first principles, would be received, I can assure you, with absolute rapture, being an ignis fatuus [*a delusive hope*] that mathematicians have been chasing from your age down to our own.
>
> EUCLID: I know it. But I cannot help you. Some mysterious flaw lies at the root of the subject.

In fact, it is *not* possible to deduce the fifth postulate from the others. This was demonstrated around 1830 by Nikolai Lobachevsky of Russia and János Bolyai of Hungary, who independently constructed a strange type of geometry in which the first four postulates hold but the fifth one does not.

Dodgson was aware of these "non-Euclidean geometries," but rejected them as irrelevant to the geometrical world in which we live. In 1888 he produced his volume *Curiosa Mathematica, Part I: A New Theory of Parallels*, in which he replaced the fifth postulate by a seemingly more "obvious" one, pictured in his frontispiece below. It asserts that

> the area inside the hexagon is larger than the area of any one of the six pieces that lie between the hexagon and the circle;

he later replaced the hexagon by a square. By assuming the truth of this self-evident postulate, he could prove the fifth postulate and all the other results equivalent to it.

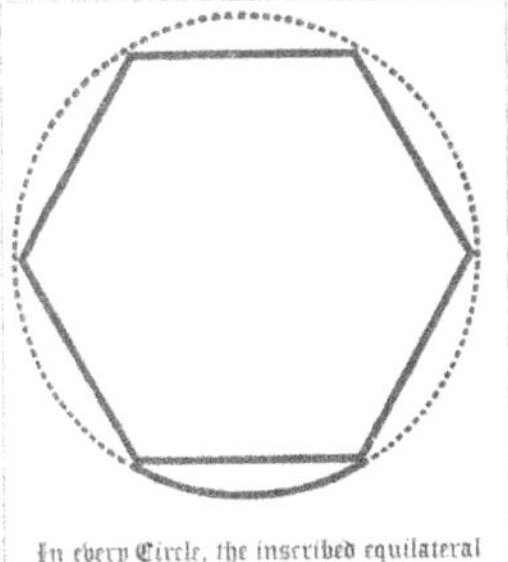

Fig. 3: Dodgson's alternative postulate

GEOMETRICAL PUZZLES AND PROBLEMS

Many of Dodgson's puzzles have a geometrical flavor. For example, in a letter to his child-friend Helen Feilden, he posed his problem of the square window:

> A gentleman (a nobleman let us say, to make it more interesting) had a sitting-room with only one window in it—a square window, 3 feet high and 3 feet wide. Now he had weak eyes, and the window gave too much light, so (don't you like "so" in a story?) he sent for the builder, and told him to alter it, so as to give half the light. Only, he was to keep it square—he was to keep it 3 feet high—and he was to keep it 3 feet wide. How did he do it? Remember, he wasn't allowed to use curtains, or shutters, or coloured glass, or anything of that sort.

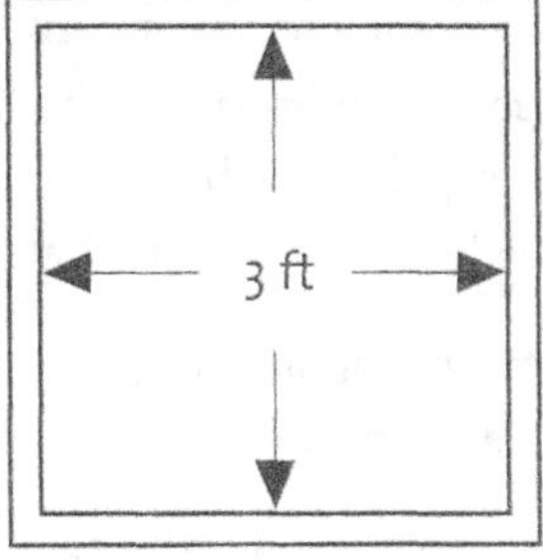

Fig. 4

A substantial collection of problems, many of them geometrical, appeared in 1893. In his Introduction to *Curiosa Mathematica II: Pillow-Problems Thought Out During Sleepless Nights*, Dodgson described how it came into being:

> Nearly all of the following seventy-two Problems are veritable "Pillow-Problems", having been solved, in the head, while lying awake at night . . . every one of them was worked out, to the very end, before drawing any diagram or writing down a single word of the solution. I generally wrote down the answer, first of all: and afterwards the question and its solution.

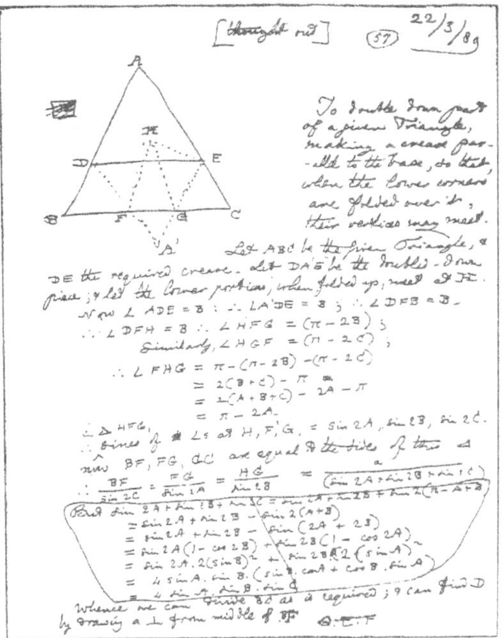
[thought out] (57) 22/3/89

To double down part of a given Triangle, making a crease parallel to the base, so that, when the lower corners are folded over it, their vertices may meet.

Let ABC be the given Triangle, & DE the required crease. Let DA'E be the doubled-down piece; & let the lower portions, when folded up, meet at H.

Now ∠ADE = B ∴ ∠A'DE = B; ∴ ∠DFB = B.

∴ ∠DFH = B ∴ ∠HFG = $(\pi - 2B)$;

Similarly, ∠HGF = $(\pi - 2C)$;

∴ ∠FHG = $\pi - (\pi - 2B) - (\pi - 2C)$
$= 2(B + C) - \pi$
$= 2(A + B + C) - 2A - \pi$
$= \pi - 2A$.

∴ sines of ∠s at H, F, G, = sin 2A, sin 2B, sin 2C.

Whence we can find BD as required; & can find D by drawing a ⊥ from middle of BF. Q.E.F.

Fig. 5: A manuscript page from *Pillow-Problems*

Many of the Pillow-Problems are highly ingenious and range through many areas of mathematics: arithmetic, algebra, pure and algebraic geometry, trigonometry, differential calculus, and probabilities. Two geometrical ones are as follows:

> 12. Given the semi-perimeter and the area of a Triangle, and also the volume of the cuboid whose edges are equal to the sides of the Triangle: find the sum of the squares of its sides.

> 49. If four equilateral Triangles be made the sides of a square Pyramid: find the ratio which its volume has to that of a Tetrahedron made of the Triangles.

RIGHT-ANGLED TRIANGLES

We conclude with the last piece of mathematics that Dodgson attempted. This was a problem that involved finding three right-angled triangles of the same area. His diary entry for December 19, 1897 reads:

> Sat up last night till 4 a.m., over a tempting problem, sent me from New York, "to find three equal rational-sided right-angled triangles. I found two, whose sides are 20, 21, 29; 12, 35, 37: but could not find three.

("Rational-sided" means that all the sides are whole numbers or fractions, while "equal" means that they have equal areas: each of Dodgson's triangles has area 210.)

In fact, he was closer than he knew. The smallest solution of this problem consists of the three right-angled triangles with sides

40, 42, 58; 24, 70, 74; and 15, 112, 113,

with common area 840; the first two of these have double the sides that Dodgson found. It is now known that there are infinitely many solutions to this problem; another is:

105, 208, 233; 91, 120, 218; and 56, 390, 394.

FURTHER READING

This chapter is adapted from parts of the author's book *Lewis Carroll in Numberland* (Penguin, 2009; and Norton, 2010), where full references and more information can be found. Also to be found in that book are Dodgson's excursion into projective geometry, his proof that all triangles are isosceles, his use of geometrical pictures in logic, and his discussion of Achilles and the tortoise couched in the language of Euclidean geometry.

Several of the mathematical pamphlets mentioned here appear in *The Mathematical Pamphlets of Charles Lutwidge Dodgson and Related Pieces*, compiled, with introductory essays, notes, and annotations, by Francine E. Abeles (Lewis Carroll Society of North America, 1994). The recreational topics mentioned here are developed in Martin Gardner's *The Universe in a Handkerchief* (Springer-Verlag, 1996) and Edward Wakeling's *Lewis Carroll's Games and Puzzles* (Dover, 1992) and *Rediscovered Lewis Carroll Puzzles* (Dover, 1995).

REFERENCE

A Carrollian Bibliography

AUGUST A. IMHOLTZ, JR.

In the preface to the Lewis Carroll section of his 2008 compilation of essays, *The Jinn from Hyperspace and Other Scribblings—Both Serious and Whimsical*, Martin Gardner wrote of himself and Carroll:

> It was not until my late twenties that I discovered the greatness of Lewis Carroll's novels about Alice's adventures in wonderland and behind the looking-glass, and how many interests I shared with the author. Like Carroll I am fond of recreational mathematics and logic, of puzzles, wordplay, magic, chess, and even aspects of theology.
>
> The more I explored books about Carroll, the more I became convinced that his *Alice* books swarmed with subtleties, jokes, and literary allusions that could be grasped by Victorian British readers but would be completely missed by today's American children. . . . I approached editors of several top publishing houses with the suggestion that they ask Bertrand Russell, known to be a great admirer of Carroll, to annotate

August A. Imholtz, Jr. is a former president of the LCSNA, and a member of its nominating committee from 1994 to date. He has spent his working life in publishing and for the past eight years has been a vice president of Readex, a digital publishing company. He has written many articles on Carroll for the Knight Letter, *the journals of both the British and the Japanese Lewis Carroll Societies, and other academic and popular publications.*

> the two *Alice* books. One editor actually wrote to Russell with the proposal, but he declined. It was Clarkson Potter, then a young editor at Crown, who suggested I do the job myself. The rest is history.

One other thing Martin Gardner had in common with Lewis Carroll is the vast number of publications he produced during his long life (1914 – 2010). Gardner wrote some 124 books and many hundreds of articles. Prof. Dana Richards, David Singmaster, and others have produced bibliographies covering his massive oeuvre, which this brief essay seeks only to complement by focusing on Martin Gardner's primary and secondary works about Lewis Carroll and on references to Carroll and his works in greater or lesser detail.

This bibliographic essay is divided into four parts. Part I covers *The Annotated Alice*, Gardner's brilliant work which was first published in 1960 and did more than anything else to place Carroll before the general populace and to establish him firmly in the academic outposts of English studies, mathematics, logic, and, of course, nonsense. Part II covers the other Gardner books on Carroll's works and references to Carroll. Part III lists, with annotations, Gardner's articles in *Scientific American* that touch on Carroll. Part IV attempts to cover, but certainly will not completely succeed, Martin Gardner's vast body of articles in periodicals other than *Scientific American* dealing, again in more or less detail, with Lewis Carroll and his works.

Martin Gardner often reprinted his separate pieces in various anthologies of his works issued by many different publishers, sometimes with different titles, a practice which can cause problems for his bibliographers. In the preface to his composite volume of essays, *Order and Surprise* (1983), he says in justification for the reprints, should any justification be necessary:

> Here and there I have modified original texts, sometimes restoring passages removed by editors, sometimes adding new footnotes to update the material. . . . It takes, I must admit, a considerable ego to put together a collection of previously published essays for a new book. I can only say that few things give a writer more satisfaction than a chance to reprint fugitive earlier scribblings, if for no other reason than to correct those inevitable copy changes made by editors and absent-minded printers.

Thanks are due to Byron Sewell and Clare Imholtz for permission to reprint their exhaustive treatment of *The Annotated Alice* and its revised volume, *The Definitive Annotated Alice* (2000), from their *Annotated International Bibliography of Lewis Carroll's Sylvie and Bruno* (Oak Knoll, 2007); to Selwyn Goodacre for notes on the British editions of many of Gardner's works; especially to Prof. Dana Richards for many corrections and additions; and to the online published bibliographies of and tributes to Martin Gardner.

THE ANNOTATED ALICES

The Annotated Alice

The Annotated Alice. New York: Clarkson N. Potter, Inc., 1960. 1–6 7–16 18–20 21–351 352, blank page Tan cloth, lettered (The / ANNOTATED / Alice) in grayish-blue. Pictorial (Tenniel's illus. of Alice with the DRINK ME bottle, in an oval, all in an overall pattern of 8-pointed devices [flowers or stars?]), printing in brown. Edges plain. Pictorial (Tenniel's illus. of Alice with the DRINK ME bottle, in an oval) blue dust jacket, printing and lettered in black, white, and gold. Back of dust jacket has a photograph of Martin Gardner in front of the Alice statue in New York's Central Park. Price $10.00 (on front dust jacket flap). NB: Clarkson Potter reprints are all dated 1960 (the first ed. date), but the actual reprinting dates are unknown. A letter code [(A), (B), (C), (D), (E) . . .] typically appears on title page verso. This likely indicates reprinting sequence: (A) = 1st reprinting; (B) = 2nd; (C) = 3rd; etc. However, the accuracy of this interpretation is uncertain

The reprinting history of this influential book is complex and identification of the various reprints is confusing. The following list is suggested, although its accuracy is uncertain. Book dealers often claim that the offered copy is a first printing, when in fact it often is not (almost all copies state 1960, probably leading to much of the confusion).

Reprints

Clarkson Potter

1960 [but later printing; 1st reprinting?]; Code (A). 1960 [but later printing; 2nd reprinting?]; Code (B). 1960 [but later printing; 3rd reprinting?]; Code (C). 1960 [but later printing; 4th reprinting?; ca. 1966, per inscription in the copy examined]; Code (D). 1960 [but later printing; 5th reprinting?]; Code (E). 1960 [but later printing; 6th reprinting?]; Code (F). 1960 [but later printing; 7th reprinting?]; Code (G). NB: This is the highest reprinting code noted.

Bramhall House

A reprinting code appears on title page verso of at least some (and probably all) Bramhall House reprints. This is in the form of an alphabetic series (ABCDE . . .), with the current reprinting being indicated by the first letter (e.g., for the sequence DEFGH, this would be the "D" [presumably 4th] Bramhall House printing). Some Bramhall House reprintings have paper covered boards and some are bound identically to the original edition or version: NY: Bramhall House, a division of Clarkson N. Potter, MCMLX [on t.p. verso]; paper covered boards.; tan cloth spine; pictorial end papers and dust jacket. NB: This is the presumed first Bramhall House reprinting. Another version: "B" [presumed 2nd] Bramhall House printing . . . (etc.): "K" [presumed 11th] Bramhall House printing; alphabetic series is: KLMNO; bound i.e., tan cloth, lettered in grayish-blue, including Potter device and name on foot of spine); title page and dust jacket front flap have Bramhall House device and name; dust jacket does not have price; dust jacket has code CJE (meaning unknown). NB: This appears to be a "hybrid" edition. There are several possible explanations, e.g., the copy examined by the Editors may have a dust jacket from a later printing. However, the book itself has aspects of both Potter and Bramhall House edition. Another version: "S" [presumed 19th] Bramhall House printing; alphabetic series is: STUVWXYZ . . . Another version: Stated 21st [presumed "U"] Bramhall House printing. Stated "AA" [presumed 27th] Bramhall House printing. NB: This is the latest Bramhall House reprinting noted.

Bookcraftsmen

New York: Bookcraftsmen Associates, 1960 [though this is probably a later reprinting].

Blond

London: A. Blond, 1960 [though this is probably a later reprinting]. Review: *TLS*, Dec. 24, 1964: 1164.

World (Forum Books)

Cleveland: World Publishing Co., [Oct] 1963; hardbound. Also issued as: Cleveland: Forum Books, Aug 1963 (1st Forum printing). Wraps. Mar 1964 (2nd printing). Unknown date (3rd printing). Nov 1965 (4th printing). Unknown date (5th printing). Unknown date (6th printing). Unknown date (7th printing). July 1969 (8th printing). May 1970 (9th printing).

Penguin

Harmondsworth, England: Penguin Books Ltd., 1965; 1–6 7–16 17–19 20, blank 21–352. Pictorial (Tenniel illus. on white background) wraps; black back wrap; orange sp. 1966. Harmondsworth and Baltimore: Penguin Books, 1970 (revised edition). 1971; yellow pictorial (Tenniel's illus. of Alice with the DRINK ME bottle) wraps, lettering and printing in dark brown. 1972. Also 1974 and 1975.

New American Library (Meridian)

New York and Scarborough, Ontario, Canada: New American Library / Times Mirror, May 1974. "A Meridian Book." White pictorial (Tenniel's illus. of Alice and the shower of cards) wraps, printing in yellow, orange, vivid pink, and black. N.B. Title page verso states: "First Printing / New American Library, May 1974" and "First Printing / World Publishing Company, October, 1963."

Macintosh Electronic Version

The Complete Annotated Alice. Santa Monica, CA: Voyager Expanded Books, 1991. "First printing December 1991". One 1.4 MB high-density floppy in a pictorial case (White Rabbit, after Tenniel). Description by Macintosh Systems. N.B. Includes *The Annotated Alice* and *More Annotated Alice*.

Wing

New York: Wing Books and Avenel, 1993; boards.; blue dust jacket. [According to Prof. Richards, the Preface to 1995 Edition lists the 17 pages which contain changes to some annotations.]

Translations

According to Dana Richards (private correspondence with Byron W. Sewell in 2003 and later correspondence with this author), *The Annotated Alice* had been translated into seven languages as of 2003, including German, Hebrew, Italian, Japanese, Portuguese, Russian, and Spanish. Gardner's Introduction may not be included in all translations (for example, the Hebrew lacks it), but for completeness, the following lists all known translations plus a few since 2003:

Dutch

Matsier, Nicolaas (trans.). *De Avonturen van Alice in Wonderland en Achter de Spiegel en Wat Alice Daar Aantrof: Aantekunigen bij Alice*. Amsterdam: Athenaeum-Polak & Van Gennep, 2009. 207 p. Boxed edition.

German

Flemming, Günther (trans.). *Alles über Alice / Alices Abenteuer im Wunderland & Durch den Spiegel und Was Alice Dort Fand*. Hamburg: Europa Verlag GmbH, August 2002. Pp. 340+. N.B.: Includes: "Die Bleistiftentwürfe von Tenniel," "Eine Anmerkung zu Lewis-Carroll-Gesellschaften," "Ausgewählte Literaturhinweise," and "Alice auf Leinwand und Bildschirm" (von David Schaefer).

Hebrew

Litvin, Rina (trans.). [*Alice in Wonderland*]. Tel Aviv: United Kibbutz Publishing House, 1997. "The New Library series," No. 13.

Litvin, Rina (trans.). [*Through the Looking-Glass*]. Tel Aviv: United Kibbutz Publishing House, 1999. [Includes the Wasp in a Wig episode.]

Italian

d'Amico, Masolino (trans.). *Alice: Le Avventure di Alice nel Paese delle Meraviglie & Attraverso lo Specchio e Quello Che Alice Vi Trovò*. Milano: Longanesi, 1971/1984.

Japanese

Issued in two vol., as follows:

Yagawa, Sumiko (trans.). *Fushigi no Kuni no Arisu* [Annotated Alice in Wonderland]. Tokyo: Tokyotosho, 1980.

Takayama, Hiroshi (trans.). *Kagami no Kuni no Arisu* [Annotated Through the Looking-Glass]. Tokyo: Tokyotosho, 1980.

Korean

Chong-min Ch'oe (trans.). *Koul nara ui Aellisu*. Soul-si: Nara Sarang, 1992. 249 p.

Portuguese

Borges, Maria Luiza X. de A. (trans.). *Alice Edição Comentada: Aventuras de Alice no País das Maravilhas & Através do Espelho*. Rio de Janeiro: Jorge Zahar, 2002.

Russian

Demurova, N[ina] M[ikhailovna] (trans.). *Priklyucheniya Alisy v Strane Chudes – Skvoz' Zerkalo i Chto Tam Uuvidela Alisa, ili Alisa v Zazerkal'e [AAiW / TtL-G and What Alice Found There, or Alice Behind the Mirror]*. Moscow: Nauka, 1978. 1–4 + port. + 5–8 9–358 359 p. Dark green cloth, lettered in gold. Pictorial dust jacket. Reprinted 1990; 1–8 9–358 359 I–VI VII, two short essays + 2 unnumbered pp. of bibliographical information; with pictorial dust jacket as above and white glazed boards.

Spanish

Oliver Torres, Francisco (trans.). *Alicia Anotada: Alicia en el País de las Maravillas & A Través del Espejo*. Madrid: Akal, 1998.

More Annotated Alice

More Annotated Alice: Alice's Adventures in Wonderland and Through the Looking-Glass and What Alice Found There by Lewis Carroll. Illustrations by Peter Newell with Notes by Martin Gardner. New York: Random House, 1990. xxxiii+363 p. Tan boards, spine lettered in gold. Bluish-green pictorial dust jacket. ["The definitive text in a new fully annotated edition." Pages xx–xxxiii consist of "*Alice's Adventures in Wonderland* from an artist's standpoint" by Peter Newell (reprinted from *Harper's Magazine*, Oct. 1901, Vol. CIII, No. DCXVII, pp. 712–717); and an original essay, "Peter Newell (1862–1924)" by Michael Patrick Hearn. Reprints (pp. 326–363) the text of *The Wasp in a Wig: A "Suppressed" Episode of Through the Looking-Glass and What Alice Found There* published by the Lewis Carroll Society of North America, see below.]

Translations

Japanese

Issued in two vol., as follows:

Takayama, Hiroshi (trans.). *Shinchu Fushigi no Kuni no Arisu*. [More Annotated Alice in Wonderland]. Tokyo: Tokyotosho, 1994.

Takayama, Hiroshi (trans.). *Shinchu Kagami no Kuni no Arisu*. [More Annotated Through the Looking-Glass]. Tokyo: Tokyotosho, 1994.

The Definitive Edition

The Annotated Alice: The Definitive Edition. New York, London: W. W. Norton & Company, 2000. Blank leaf + i–vi vii viii, blank xi x, blank xi–xxviii 1–3 4, blank 5 6 7– 312 + 5 blank leaves. Red boards, blocked in gold. Spine lettered in gold. Dark brownish-red pictorial dust jacket, lettered in gold. NB: Reprints Gardner's original introduction on pp. xiii–xxii. Wraps. Also Harmondsworth: Penguin, 2001. And 2nd printing (includes numerical string 234567890 on title page verso).

CARROLL AND OTHER BOOKS

Alice's Adventures in Wonderland

Alice's Adventures in Wonderland & Through the Looking-Glass. With a new introduction and annotated bibliography by Martin Gardner. New York : Signet Classic, 2000. x+239 p. [Introduction reprinted in *The Jinn from Hyperspace and Other Scribblings – Both Serious and Whimsical*. Amherst, New York: Prometheus Books, 2008. Pp. 301–307.]

The Ambidextrous Universe

The Ambidextrous Universe. Illustrated by John Mackey. New York: Basic Books, 1964. x + 294 p. Revised as *The Ambidextrous Universe: Left, Right and the Fall of Parity*. Illustrated by John Mackey. New York: New American Library, 1969. 254 p. (A Mentor Book). Reprinted as *The Ambidextrous Universe: Mirror Symmetry and Time-Reversed Worlds*. New York: Charles Scribner's Sons, 1979. [Discusses mirror imagery in the *Alice* books (pp. 5, 69, 114) and the outlandish watch with its "reverse peg" in *Sylvie and Bruno Concluded* (p. 266).] Later issued as *The New Ambidextrous Universe: Symmetry and Asymmetry from Mirror Reflections to Superstrings*. New York: W. H. Freeman, 1990. xiv + 392 p.

Translations

French

Roux, Claude (trans.) *L'univers ambidextre: la droite, la gauche et la faillite de la parité*. Paris: Dunod, 1968. iv–266 p.

Roux, Claude and Alain Laverne (trans.). *L'univers ambidextre: les miroirs de l'espace-temps*. Nouv. éd., rev. et mise à jour / par l'auteur Paris: Éd. du Seuil, 1994. Impression: 45-Manchecourt; Impr. Maury eurolivres. 357 p.

Italian

Luca, Speranza and Pedrazzini Guido (trans.). *L'universo ambidestro: nel mondo degli specchi, delle asimmetrie, delle inversioni temporali*. Bologna: Zanichelli, 1984. Repr. Novara: Mondadori-De Agostini, 1996.

Spanish

Freixenet, Joan Tarrés (trans.). El universo ambidiestro. Barcelona: RBA, D. L. 1994.

Alice's Adventures under Ground

Alice's Adventures under Ground by Lewis Carroll. Facsimile of the author's manuscript book with additional material from the facsimile edition of 1886. With a new introduction by Martin Gardner. New York: Dover Publications Inc., 1965. xi+90+[19] p. Paperback with decorated front and rear covers. First edition with price of $1.00. Subsequent Dover editions have prices of $1.50, $2.00, etc. [Introduction reprinted in *The Jinn* (op. cit.).] Hardcopy reprint in dust wrapper issued by McGraw Hill, New York, 1966. First edition, As above, but in red cloth boards. Gloucester, MA: Peter Smith, 1987.

Alice in Puzzle-Land

Alice in Puzzle-Land, A Carrollian Tale for Children Under Eighty, by Raymond M. Smullyan, with an introduction by Martin Gardner, illustrated by Greer Fitting. New York: William Morrow and Company, Inc., 1982. x +182 p. First edition, in dust wrapper. First British edition: (paperback) Harmondsworth, U.K.: Penguin Books, 1984.

Translations

German

Brandt, Thea (trans.). *Alice im Rätselland: phantastische Rätselgeschichten, abenteuerliche Fangfragen und logische Traumreisen*. Frankfurt (Main): Krüger, 1984. 205 p.

Portuguese

Ribeiro, Vera (trans.). *Alice no País dos Enigmas*. Rio de Janeiro: J. Zahar Ed., 2003. 191 p.

Russian

Jakopin, Gitica (trans.). *Alisa v strane smekalki*. Moscow: Mir, 1987. 179 + iii p.

Slovenian

Jakopin, Gitica (trans.). *Alica v deželi ugank: zgodbe za otroke pod osemdeset*. Ljubljana, Slovenia: Državna založba Slovenije, 1989.

Spanish

Millán, Montserrat (trans.). *Alicia en el país de las adivinanzas: un cuento al estilo de Lewis Carroll para niños menores de ochenta años*. Madrid: Ca_tedra, 1984. 207 p.

The Annotated Snark

The Annotated Snark: The full text of Lewis Carroll's great nonsense epic The Hunting of the Snark *and the original illustrations by Henry Holiday. With an introduction and notes by Martin Gardner.* New York: Simon and Schuster, 1962. 111 p. [Pink topstain.] Reprint as above by Bramhill House, a division of Clarkson N. Potter, Inc., 1962, but no pink topstain. Penguin paperback editions: *The Annotated Snark, with an Introduction and Notes by Martin Gardner*. Harmondsworth: Penguin Books, 1967. 125 p. First edition 1967; reprinted with revisions 1973; reprinted 1975; third reprint 1977; fourth reprint 1979; fifth reprint 1984; sixth reprint 1987; reprint "1974" code 7 9 10 8; reprint "1974" code 9 10 8. Reprinted as "Penguin Classics," 1995. code 1 3 5 7 9 10 8 6 4 2 .

Lewis Carroll's "The Hunting of the Snark," illustrated by Henry Holiday – Centennial Edition, edited by James Tanis and John Dooley. Los Altos: William Kaufmann, Inc. in cooperation with Bryn Mawr College Library, 1981. Subscriber's Edition – limited to 395 numbered copies, signed by the participants; Collector's Edition – limited to 1,995 copies [lacks facsimile of first edition of *The Hunting of the Snark*] vii+129+44 plates; Trade edition of 5000 copies (reprinted with emendations in February, 1982) as above. vii+129+xi+83+[5] p. Contains *The Annotated Snark*, among other texts.

The Annotated Hunting of the Snark. The definitive edition. The full text of Lewis Carroll's great nonsense epic The Hunting of the Snark. *Original illustrations by Henry Holiday. Edited with a preface and notes by Martin Gardner*. Introduction by Adam Gopnik. New York: W. W. Norton & Co., 2006. xli+152+[2] p.

Encyclopedia of Philosophy

Encyclopedia of Philosophy, ed. Paul Edwards. New York: Collier Macmillan, 1967. Martin Gardner's article "Logic Diagrams" discusses Carroll's *The Game of Logic* briefly. Vol. 5, pp. 77–81.

The Hunting of the Snark

The Hunting of the Snark. London: Catalpa Press / Ian Hodgkins and Co Ltd., 1974. [44] p. 1 twelve-foot foldout of the crew plus 10 rearrangeable trisected cards depicting portraits of the crew members. [Illustrated by Byron Sewell with an introduction by Martin Gardner. Limited to 250 copies.]

Knotted Doughnuts

Knotted Doughnuts and Other Mathematical Diversions. New York: W. H. Freeman, 1986. xii+278 p. [Addendum to "The gray binary code" chapter discusses Carroll's Doublets, p. 20–21.]

The Last Recreations

The Last Recreations: Hydras, Eggs, and Other Mathematical Mystifications. New York: Copernicus, 1997. x + 392 p. [Discusses the fact that Carroll was one of the last of famous mathematicians to doubt non-Euclidean geometry, in the chapter "Non-Euclidean geometry," pp. 303–315.]

Lewis Carroll's Chess Wordgame

Lewis Carroll's Chess Wordgame. Pasadena, CA: Kadon Enterprises, 1991. [A Carrollian wordgame with a cloth tablecloth chessboard and 55 letters in two colors. The 12-page booklet of rules is by Martin Gardner. Issued in a brown cardboard box, 21 × 21 cm.]

Logic Machines and Diagrams

Logic Machines and Diagrams. New York: McGraw-Hill and Company, 1958. 157 p. [Discusses Carroll's diagrammatic method for solving logical problems pp. 45–48 with further brief mentions of Carroll on pp. 51, 53, 76, 94, and 106.] Reprinted under the title *Logic Machines, Diagrams, and Boolean Algebra*. New York: Dover Publications, 1968. xi+157 p.; also reprinted under the original title with a foreword by Donald Michie. Chicago: University of Chicago Press, 1982. xiv+165 p.

Translation

French

Ghezzi , Mary (trans.). *L'étonnante histoire des machines logiques*. Paris: Dunod, 1964. 199 p.

The Night Is Large

The Night is Large: Collected Essays, 1938–1995. New York: St. Martin's Press, 1996. xix+586 p. "Coleridge and the Ancient Mariner," pp. 297–317. [Discusses Carroll's "You are old, Father William" parody of Southey's poem.] "Lewis Carroll and His Alice Books," pp. 318–327. [Reprints the opening chapter of *The Annotated Alice* with a postscript correcting four errors.]

The Nursery Alice

The Nursery Alice. By Lewis Carroll with a new introduction by Martin Gardner. New York: Dover Publications, Inc., 1966. xi+56+[7] p. First edition [?] $2.50, later edition $4.95. [Introduction reprinted in *The Jinn* (op. cit.), pp. 287–292.] Hardcopy reprint in dust wrapper of the above Dover paperback edition issued by McGraw-Hill, New York, 1966.

Translations

Spanish

Gervás, Agustín (trans.). *Alicia para los pequeños*. [Madrid]: Alfaguara [1977]. 66 p. [epilogue by Martin Gardner].

Phantasmagoria

Phantasmagoria. Introduction and notes by Martin Gardner. Illustrations by Arthur B. Frost. Amherst, New York: Prometheus Books, 1998. 73 p. [Introduction reprinted in *The Jinn* (op. cit.), pp. 281–286.]

The Snark Puzzle Book

The Snark Puzzle Book by Martin Gardner. New York: Simon and Schuster, 1973. 124 p. First edition code 1 2 3 4 5 6 7 8 9 10. In dust wrapper. Reprinted, New York: Prometheus Books, 1990. First edition thus, code 1 2 3 4 5 6 7 8 9 10. In dust wrapper.

Sylvie and Bruno

Sylvie and Bruno by Lewis Carroll. With 46 original illustrations by Harry Furniss and a new introduction by Martin Gardner. New York: Dover Publications, Inc., 1988. xxxi+400 p. [Reprinted in *The Jinn* (op. cit.), pp. 265–280.]

The Universe in a Handkerchief

The Universe in a Handkerchief. Lewis Carroll's Mathematical Recreations, Games, Puzzles, and Word Plays, by Martin Gardner. New York: Copernicus, 1996. x+158 p. First edition in dust wrapper.

Translations

Japanese

Momma, Yoshiyuki and Naoko Momma (trans.). *Ruisu kyaroru asobi no uchu.* Tokyo: Hakuyosha, 1998. 253 p.

Korean

Jink Won Kim (trans.). [Mathematician in Wonderland]. Seoul: Purun Media, 2000.

Polish

Bartol, Wiktor (trans.). *Wszechswiat w chusteczce: Rozrywki matematyczne, a takze zabawy, łamigłówki i gry słowne Lewisa Carrolla.* Warsaw: Prószynski i S-ka, 1999. 203 p.

Visitors from Oz

Visitors from Oz. New York: St. Martin's Press. 1998. xvi+189 p. [A work of fiction, chapters 6 through 10 ("Wonderland," "The Mad Hatter," "Humpty Dumpty," "The White Knight," and "The Red King") involve figures from the *Alice* books.]

The Wasp in a Wig

The Wasp in a Wig, a "Suppressed" Episode of Through the Looking-Glass and What Alice Found There, *with a Preface, Introduction and Notes by Martin Gardner.* New York: The Lewis Carroll Society of North America, 1977. xiv+21+foldout facsimile of galley proofs [slips 64–67 and portions of 63 and 68] +[2] p. Carroll Studies No. 2. 500 copies. Limited edition. Red paper wraps. [Martin Gardner was instrumental in persuading Norman Armour, Jr., the owner of the "Wasp in a Wig" galley sheets, to allow the Lewis Carroll Society of North America to publish the episode.]

Editions

Cloth bound issue, as above but with three stamped gold borders on cover. 750 copies.

First English edition: London: Macmillan, 1977. Frontispiece by Ken Leeder, in dust wrapper, with the Ken Leeder illustration on the front.

First American trade edition: Clarkson N. Potter, New York, 1977. In dust wrapper. Dust wrapper says "Frontispiece after Tenniel with two additional black-and-white illustrations by Ralph Steadman," but no such illustrations are to be found. As above, but facsimiles of galley slips do not fold out.

American trade edition: Clarkson N. Potter, New York, second printing, 1978 (noted on title page reverse). In dust wrapper. Dust wrapper says "Frontispiece after Tenniel with two additional black-and-white illustrations by Ralph Steadman," but the book contains only the frontispiece and *one* extra illustration. As above, but facsimiles of galley slips do not fold out.

American trade edition: Clarkson N. Potter, New York, third printing, undated. Only indication of "third printing" is on front flap of the dust wrapper. Dust wrapper also says it has "Frontispiece (after Tenniel) and an additional illustration by Ralph Steadman," but, again, none is to be found. As above, but facsimiles of galley slips do not fold out.

Translations

French

Roy, Bruno (trans.). *Le frelon a perruque: un chapitre inédit de L'autre côté du miroir*. Montpellier: Bibliothèque et litteraire, 1979. Anne-Marie Soulcié, illus. 46 p.

Japanese

Yanase, Naoki (trans.). *Katsura o kabutta suzumebachi*. Tokyo: Rengashobo Shin Sha, 1978. 44+42 p.

Spanish

Miguel, Olivia de (trans.). *La avispa con peluca: Un episodio suprimido de A través del espejo y lo que Alicia encontró allí*. Published as an offprint to number 12 of the Catalan-language quarterly *PUB-21: la veu del Batxillerat* in Barcelona, 1981.

SCIENTIFIC AMERICAN "MATHEMATICAL GAMES" ARTICLES

Vol. 197, no. 4 (October, 1957): 130–136.
"How to remember numbers by mnemonic devices such as cuff links and red zebras"
[Mentions Lewis Carroll's "Memoria Technica."] Reprinted under the title "Memorizing Numbers" in *The Scientific American Book of Mathematical Puzzles and Diversions* (New York: Simon and Schuster, 1959), pp. 102–109; repr. in *Hexaflexagons and Other Mathematical Diversions: The First Scientific American Book of Puzzles and Games* (Chicago: University of Chicago Press, 1988); also reprinted under the title "Memorizing Numbers" in *Hexaflexagons, Probability Paradoxes, and the Tower of Hanoi: Martin Gardner's First Book of Mathematical Puzzles and Games* (The New Martin Gardner Mathematical Library), pp. 115–121.

Vol. 200, no. 2 (February, 1959): 136–140.
"Brain teasers that involve formal logic"
[Discusses the premises in Lewis Carroll's *Symbolic Logic* from which one is to deduce that "no magistrates are snuff takers."] Reprinted in *The Second Scientific American Book of Mathematical Puzzles and Diversions* (New York: Simon and Schuster, 1961), pp. 119–129; reprinted (Chicago: University of Chicago Press, 1987); and also reprinted in *Origami, Eleusis, and the Soma Cube* (Cambridge University Press, 2008), pp. 106–114, under the title "Recreational Logic," with a postscript by Martin Gardner, pp. 115–116.

Vol. 201, no. 1 (July, 1959): 138–143.
"About Origami, the Japanese art of folding objects out of paper"
[Mentions Lewis Carroll's interest in paper folding.] Reprinted in *The Second Book* (op. cit.); and also reprinted in *Origami* (op. cit.), pp. 160–163, under the title "Origami."

Vol. 202, no. 3 (March, 1960): 172–182.
"The games and puzzles of Lewis Carroll, and answers to February's problems"
[Discusses in some detail a number of Lewis Carroll's games and puzzles. Reproduces the famous self-assisted photograph circa July 1876.] Reprinted under the title "The Games and Puzzles of Lewis Carroll" in *New Mathematical Diversions from Scientific American* (New York: Simon and Schuster, 1966) pp. 47–57; reprinted (London: Allen and Unwin, 1969); and (Chicago: University of Chicago Press, 1983), pp. 178–184.

Vol. 204, no. 2 (February, 1961): 146–153.
"Diversions that involve one of the classic conic sections: the ellipse"
[Mentions Lewis Carroll's pamphlet about a circular billiard table.] Reprinted in *New Mathematical Diversions* (op. cit.); in Morris Kline's *Mathematics: An Introduction to Its Spirit and Use* (New York: W. H. Freeman, 1979), pp. 100–103; also in *Sphere Packing, Lewis Carroll, and Reversi* (Cambridge: Cambridge University Press The New Martin Gardner Mathematical Library, 2009), pp. 189–201, under the title "The Ellipse."

Vol. 209, no. 4 (September, 1963): 248–265.
"How to solve puzzles by graphing rebounds of a bouncing ball"
[Mentions Lewis Carroll's pamphlet "Circular Billiards."] Reprinted under the title "Bouncing Balls in Polygons and Polyhedrons" in *Martin Gardner's Sixth Book of Mathematical Diversions from Scientific American* (San Francisco: W. H. Freeman, 1971), pp. 29–36; reprinted (Chicago: University of Chicago Press, 1983).

Vol. 210, no. 4 (April, 1964): 126–135.
"Various problems based on planar graphs, or sets of 'vertices' and 'edges'"
[Mentions Carroll urging young girls to construct Euler graphs.] Reprinted in *Mathematics: An Introduction* (op. cit.), pp. 125–129.

Vol. 211, no. 4 (September, 1964): 218–224.
"Puns, palindromes and other word games that partake of the mathematical spirit"
[Discusses Lewis Carroll and James Joyce.] Reprinted under the title "Word Play" in the *Sixth Book* (op. cit.), pp. 143–151.

Vol. 212, no. 4 (April, 1965): 128–135.
"The infinite regress in philosophy, literature and mathematical proof"
[Mentions Lewis Carroll's "What the Tortoise said to Achilles."] Reprinted under the title "Infinite Regression" in the *Sixth Book* (op. cit.), pp. 220–229.

Vol. 215, no. 2 (August, 1966): 96–99.
"Puzzles that can be solved by reasoning on elementary physical principles"
[Discusses Mein Herr's idea of oval wheels on a carriage in *Sylvie and Bruno Concluded*, chapter 7, and converts the idea into a problem.] Reprinted in *Mathematical Carnival* (New York: Knopf, 1975; repr. Washington D.C.:

Mathematical Association of America, 1989) under the title "The Rising Hourglass and Other Physics Problems," pp. 173–193.

Vol. 215, no. 3 (September, 1966): 264–272.
"The problem of Mrs. Perkins quilt, and answers to last month's puzzles"
[Discusses answer to Lewis Carroll's "carriage" problem posed in the August column.]

Vol. 215, no. 5 (November, 1966): 41–54.
"Is it possible to visualize a four-dimensional figure?"
[Quotes from Lewis Padgett's short story "Mimsy Were the Borogoves" and refers to "Jabberwocky."] Reprinted in *Mathematical Carnival* (op. cit.) as "Hypercubes."

Vol. 216, no. 1 (January, 1967): 118–123.
"Dr. Matrix delivers a talk on acrostics"
[Discusses Carroll's acrostics.]

Vol. 216, no. 5. (May, 1967): 135–141.
"Cube root extraction and the calendar trick, or how to cheat in mathematics"
[Discusses Carroll's method for calculating the day of the week.]

Vol. 217, no. 3 (September, 1967): 268–276.
"Double acrostics, stylized Victorian ancestors of today's crossword puzzles"
[Discusses Carroll's acrostics, especially his most difficult one, which is found in the 1869 version of *Phantasmagoria*.] Reprinted in *Mathematical Magic Show: More Puzzles, Games, Diversions, Illusions & Other Mathematical Sleight-of-mind from* Scientific American*; With Repartee from Readers; Afterthoughts from the Author* (New York: Knopf, 1967), pp. 86–93; reprinted (New York: Vintage Books, 1971).

Vol. 218, no. 3 (February, 1968): 118–123.
"Combinatorial problems involving 'tree' graphs and forests of trees"
[Discusses ambiguity in Lewis Carroll's double acrostic ballad (see September 1967 column).] Reprinted under the title "Trees" in *Magic Show* (op. cit.).

Vol. 220, no. 4 (April, 1969): 124–126.
"An octet of problems that emphasize gamesmanship, logic and probability"
[Discusses Carroll's Fifth Pillow-Problem.] Reprinted under the title "The Rotating Round Table and Other Problems" in *Mathematical Circus* (New York: Knopf, 1980), pp. 182–202.

Vol. 224, no. 4 (April, 1971): 114–117.
"Geometric fallacies: hidden errors pave the road to absurd conclusions"
[Discusses two theorems from *Pillow-Problems*: Theorem 1 – An obtuse angle is sometimes equal to a right angle; Theorem 2 – Every triangle is isosceles.] Reprinted in *Mathematics: An Introduction* (op. cit.), pp. 118–12; also in *Wheels, Life and Other Mathematical Amusements* (New York: W. H. Freeman, 1983), pp. 51–60.

Vol. 224, no. 5 (May, 1971): 110–116.
"The combinatorial richness of folding a piece of paper"
[Discusses explanations of the geometric fallacies discussed in the April 1971 column, which were of interest to Lewis Carroll.] Reprinted under the title "Paper Folding" in *The Colossal Book of Mathematics*, (New York: Norton, 2001).

Vol. 225, no. 6 (December, 1971): 96–99.
"Further encounters with touching cubes, and the paradoxes of Zeno as 'supertasks'"
[Discusses Lewis Carroll's "What the Tortoise Said to Achilles."] Reprinted under the title "Salmon on Austin's Dog" in *Wheels, Life* (op. cit.), pp. 134–141.

Vol. 227, no. 4 (October, 1972): 110–112.
"Why the long arm of coincidence is not as long as it seems"
[Mentions Carroll's diary observation that most good things happened to him on Tuesdays.] Reprinted under the title "Coincidence" in *Knotted Doughnuts and Other Mathematical Entertainments* (op. cit.), pp. 1–10.

Vol. 228, no. 2 (February, 1973): 106–109.
"Up-and-down elevator games and Piet Hein's mathematical puzzles"
[Discusses Lewis Carroll's failure to design a tournament system that does the best possible justice to the second-best player.] Reprinted under the title "Elevators" in *Knotted Doughnuts* (op. cit.), pp. 123–132.

Vol. 230, no. 4 (April, 1974): 110–114.
"Nine challenging problems, some rational and some not"
[Discusses a "March Hare and Mad Hatter" problem.] Reprinted under the title "Reverse the Fish and Other Problems" in *Knotted Doughnuts* (op. cit.), pp. 176–191.

Vol. 231, no. 2 (August, 1974): 98–103.
"On the fanciful history and creative challenges of the puzzle game of tangrams"
[Discusses *The Fashionable Chinese Puzzle*, one of the earliest English books on tangrams – Lewis Carroll's copy of which was acquired by the English mathematician and puzzle creator Henry Ernest Dudeney.] Reprinted under the title "Tangrams, Part I" in *Time Travel and Other Mathematical Bewilderments* (New York: W. H. Freeman, 1988), pp. 27–37.

Vol. 231, no. 4 (October, 1974): 120–125.
"On the paradoxical situations that arise from nontransitive relations"
[Discusses the Condorcet effect paradox, which Lewis Carroll rediscovered in his pamphlets on voting.] Reprinted under the title "Combinational Theory" in the *Sixth Book* (op. cit.), pp. 18–27; also reprinted in *Time Travel* (op. cit.), pp. 55–69.

Vol. 232, no. 2 (February, 1975): 98–102.
"How the absence of anything leads to thoughts of nothing"
[Discusses the idea of "nothing" in the *Alice* books and reprints the frontispiece of Ferdinand Canning Scott Schiller's Special Christmas, 1901, parody of the British philosophy journal *Mind*.] Reprinted in *Magic Show* (op. cit.), pp. 278–283; also reprinted in *The Night Is Large* (op. cit.), pp. 397–411.

Vol. 233, no. 5 (November 1975): 120–125.
"On map projections (with special reference to some inspired ones)"
[Quotes map verses from *The Hunting of the Snark*.] Reprinted under the title "The Sixth Symbol and Other Problems" in *Time Travel* (op. cit.), pp. 205–212.

Vol. 233, no. 6 (December, 1975): 116–119.
"A random assortment of puzzles, together with reader response to earlier problems".
[Discusses Lewis Carroll's "anagrammatic sonnet" to Maud Stanton.] Reprinted in *Time Travel* (op. cit.), pp. 205–210.

Vol. 234, no. 1 (January, 1976): 118–127.
"A breakthrough in magic squares, and the first perfect magic cube"
[Discusses solutions, including the first one published, to Lewis Carroll's anagrammatic sonnet (see December, 1975 column).] Reprinted under the title "Magic Squares and Cubes" in *Time Travel* (op. cit.), pp. 213–226.

Vol. 234, no. 4 (April, 1976): 126–130.
"Snarks, Boojums, and other conjectures related to the four-color-map theorem".
[Proposes calling nontrivial uncolorable trivalent graphs "Snarks."] Reprinted under the title "Trivalent Graphs, Snarks, and Boojums" in *The Last Recreations* (op. cit.), pp. 361–380.

Vol. 237, no. 3 (September, 1977): 24–42.
"On conic sections, ruled surfaces and other manifestations of the hyperbola" [Quotes the hyperbola passage from "The Dynamics of a Particle."] Reprinted under the title "Hyperbolas" in *Penrose Tiles and Trapdoor Ciphers* (New York: W. H. Freeman, 1989) pp. 205–218.

Vol. 238, no. 3 (March, 1978): 25–29.
"Count Dracula, Alice, Portia, and many others consider various twists of logic" [Discusses logic problems encountered by Alice in the Forest of Forgetfulness in *Through the Looking-Glass*.] Reprinted under the title "Raymond Smullyan's Logic Puzzles" in *Penrose Tiles* (op. cit), pp. 281–292.

Vol. 239, no. 5 (November, 1978): 22–33.
"In which a mathematical aesthetic is applied to modern minimal art"
[Discusses "The New Belfry of Christ Church, Oxford" and Lewis Carroll's view that the ordinary cube is too minimal to have any aesthetic value.] Reprinted with additional material in *Fractal Music, Hypercards and More* (New York: W. H. Freeman and Company, 1992) under the title "Minimal Sculpture."

Vol. 241, no. 1 (July, 1979): 16–24.
"Douglas R. Hofstadter's *Gödel, Escher, Bach*"
[Review of Hofstadter's book with a discussion of Carroll's "Achilles and the Tortoise" paradox.] Reprinted in *Fractal Music* (op. cit.).

Vol. 241, no. 3 (September, 1979): 22–32
"In some patterns of numbers or words there may be less than meets the eye" [Mentions Carroll's letter to children that appear to be prose but are actually verse.] Reprinted under the title "Pi and Poetry: Some Accidental Patterns" in *Fractal Music* (op. cit), pp. 271–280.

Vol. 241, no. 4 (October, 1979): 18–26.
"Some packing problems that cannot be solved by sitting on the suitcase"
[Quotes "It *is* a Square . . . Beautiful! Beau-ti-ful. Equilateral. *And* Rectangular" from *A Tangled Tale*, Knot 2.] Reprinted in *Fractal Music* (op. cit.), p. 289.

Vol. 245, no. 4 (October, 1981): 23–34.
"Euclid's parallel postulate and its modern offspring"
[Mentions Carroll being one of the last mathematicians to doubt non-Euclidean geometry.] Reprinted under the title "Non-Euclidean geometry" in *The Last Recreations* (op. cit.), pp. 303–316.

PERIODICALS OTHER THAN *SCIENTIFIC AMERICAN*

Bandersnatch: The Newsletter of the Lewis Carroll Society (UK)

Issue 128 (July 2005) [Letter to the editor]
[Discusses Carroll and Baum, pointing out that the first word in *Alice's Adventures in Wonderland* is "Alice" and the first word in *The Wizard of Oz* is "Dorothy."]

The Carrollian: The Lewis Carroll Journal (The Lewis Carroll Society [UK]*)*

No. 13 (Spring 2004): 57. "Anne Thackeray's *From an Island*"
[Discusses whether Lewis Carroll is a character in Thackeray's novel, as Karoline Leach suggested.]

Chicago Tribune Book World

Vol. 127, no. 287 (October 14, 1973): 9. "The Magic of Lewis Carroll"
[Review of John Fisher's *The Magic of Lewis Carroll.*] Reprinted in *Order and Surprise* (Buffalo: Prometheus Books, 1983), pp. 289–291. Also printed in *The Washington Post*, see below.

Vol. 132, no. 296 (October 23, 1978): 9.
[Review of Roger Sale's *Fairy Tales and After* with discussion of Sale's essay on Carroll.] Reprinted in *Order and Surprise* (op. cit.), pp. 326–327, under the title "The Glory of Everything."

Children's Literature

Vol. 18 (1990): 145–159.
[Includes Martin Gardner's "*John Martin's Book*: An Almost Forgotten Children's Magazine"; with brief comparison of Carroll and Martin.] Reprinted in *From the Wandering Jew to William F. Buckley, Jr.* (Amherst, New York: Prometheus Books, 2000), pp. 77–96.

The Magazine of Fantasy and Science Fiction

Vol. 8, No. 1 & No. 2 (January & February 1955). "The Royal Historian of Oz" [Discusses Baum and Carroll.] Reprinted in *Order and Surprise* (op. cit.), pp. 115–138. Also reprinted and expanded in *The Night Is Large* (op. cit.), pp. 328–247.

Jabberwocky, The Journal of the Lewis Carroll Society (UK)

Vol. 14, no. 3 (Summer 1985): 63–64. "Who wrote 'Speak Gently'? – An update"

Vol. 21, no. 2 (Spring 1992): 55. "An unrecorded board game with reference to Jeffrey Stern's article in *Jabberwocky*"
[The title of Stern's article is "Carroll identifies himself at last, or: A problem solved and a puzzle posed."]

Knight Letter, The Magazine of the Lewis Carroll Society of North America

Vol. I, No. 57 (Spring 1998): 10. [Letter to the editor]
[Discusses a problem posed by Henry Ernest Dudeney involving three triangles, one or more with sides in rational fractions.]

Vol I, No. 65 (Winter 2000): 7. "Well, you know, . . . "
[Discusses the interjected phrase "Well, you know," in the *Alice* books.] Reprinted in the present volume.

Vol. II, iss. 5, no. 75 (Summer 2005): 1–7. "A Gardner's Bouquet: New Annotations"
[Additional annotations and corrections for *The Annotated Alice*.] Reprinted in the present volume.

Vol. II, iss. 6, no. 76 (Spring 2006): 1–3. "A Gardner's Nosegay: Further Annotations"
[Additional annotations and corrections for *The Annotated Alice*.] Reprinted in the present volume.

Los Angeles Times

December 6, 1998. "Breathless," pp. 2.
[Review of *Reflections in a Looking-Glass* by Morton N. Cohen.] Reprinted in *From the Wandering Jew to William F. Buckley Jr.* (op. cit.), under the title "Lewis Carroll, Photographer," pp. 171–174.

Math Horizons

Vol. 2 (November, 1994). "Word Ladders – Lewis Carroll's Doublets"
Reprinted in *A Gardner's Workout: Training the Mind and Entertaining the Spirit* (Natick, Massachusetts: A. K. Peters, 2001), pp. 133–38.

The Mathematical Gazette

Vol. 80, no. 487 (March 1996): 195–198. "Word Ladders: Lewis Carroll's Doublets"

Nature

Vol. 379, no. 6561 (January 11, 1996): 127–128. "Morton Cohen's Life of Lewis Carroll"
[Review of *Lewis Carroll: A Biography* by Morton N. Cohen.] Reprinted in *From the Wandering Jew* (op. cit.), pp. 167–170.

New York Herald Tribune

Vol. CXXI, no.42,263 (August 19, 1962): 8."The Mathematical Magpie"
[Discusses Lewis Carroll's "cross cap" – a surface with no inside or outside – in a review of Clifton Fadiman's *The Mathematical Magpie* and Oswald Jacoby's *Mathematics for Pleasure*.] Reprinted in *Order and Surprise* (op. cit.), p. 254.

New York Review of Books

Vol. 34, no. 19 (December 3, 1987): 34–36 "Infinity and information"
[Reviews Rudy Rucker's *Mind Tools* and mentions White Queen's believing six impossible things before breakfast.] Reprinted in *The Night Is Large* (op. cit.), 50–67.

New York Times Book Review

Vol. CXXV, no. 43093 (January 18, 1976): 8 & 12. "Children's Books Were Whatever Grownups Gave to Children". [Reviews *Cobwebs to Catch Flies* by Joyce Irene Whalley and *Early Children's Books and Their Illustrators* by Gerald Gottlieb.]
[Notes that the *Alice* books were originally more popular with teenagers and grownups than with children.]

Parabola

Vol. 11, no. 2 (May 1986): 38–41. "From *The Annotated Alice*"
[Reprints two long notes from *Alice*, in a special issue devoted to "Mirrors."]

Quantum: The Magazine of Math and Science

Vol. 3, no. 6 (July/August 1993): 30–31, 60. "A Royal Problem"
[Discussion of a chess problem. Written with Andy Liu.]

Vol. 5, no. 6 (March/April 1995): 40. "Lewis Carroll's Sleepless Nights: Two Problems for Insomniacs"

[Reprinted in *Mischmasch, the Journal of the Lewis Carroll Society of Japan*. No. 2, 1997, pp. 140–143 (translated by Katsuko Kasai)]. Also reprinted in *A Gardner's Workout* (op. cit.), 129–132.]

Pediatrics: Official Journal of the American Academy of Pediatrics

Vol. 62 (August 1978): 227. "Which Way to Go"
[Includes excerpt from *The Annotated Alice*.]

Saturday Review

Vol. 48, no. 29 (July 17, 1965): 18–19. "A Child's Garden of Bewilderment: An Anniversary for Alice"
[Reprinted in *Order and Surprise* (op. cit.), pp. 164–168, under the title "Carroll versus Baum"; and in *Only Connect: Readings on Children's Literature*, ed. by Sheila Egoff, G. T. Stubbs, and L. F. Ashley (New York and Toronto: Oxford University Press, 1969; repr. 1980, 1996), pp. 150–155.]

Semiotica

Vol. 5, no. 1 (1972): 89–92. "Lewis Carroll's Semiotics"
[Review of Robert D. Sutherland's *Language and Lewis Carroll.*] Reprinted in *Order and Surprise* (op. cit.), pp. 280–283.

Skeptical Inquirer

Vol. 19, no. 4 (July/August 1995): 3–5. "Klingon and Other Artificial Languages"
[Mentions the Esperanto translations of the *Alice* books.] Reprinted in *Weird Water and Fuzzy Logic: More Notes of a Fringe Watcher* (Amherst, New York: Prometheus Books, 1996), pp. 147–154. Also reprinted in *The Night Is Large* (op. cit.), pp. 162–169.

Washington Post Book World

Vol. 17, no. 40 (October 14, 1973): 4. "Magical Mystery Tour"
[Review of *The Magic of Lewis Carroll* by John Fisher. Also printed in the *Chicago Tribune*, see above.]

Word Ways: The Journal of Recreational Linguistics

Vol. 14, no. 1 (February 1981). "Kickshaws: A Miscellany of Word Play"
[Discusses Carrollian wordplay.] Reprinted in *Order and Surprise* (op. cit.), pp. 188–194.

Acknowledgements

Naturally, a book like this involves the collected efforts of many dedicated souls. The editor would particularly like to first thank the members of our sister Lewis Carroll Societies who have enthusiastically supported our efforts to make this happen; their websites are listed below. A big shout-out to our designer, Andrew Ogus, and to our cover artist, Mahendra Singh, who have given so unselfishly of their time and talent. To our intrepid proofreaders Michael Patrick Hearn (for all his "niggling"), Sandra Parker, and Andrew Sellon. To Jim Gardner, who so generously mailed to us his photos of his father and gave us permission to scan and use them. And to my dear and wonderful family for their loving support: Llisa, Sonja, and Martin.

The Lewis Carroll Society of North America
www.LewisCarroll.org

The Lewis Carroll Society (UK)
LewisCarrollSociety.org.uk

The Lewis Carroll Society of Japan
www.soc.nii.ac.jp/lcsj

The Lewis Carroll Society of Brazil
www.lewiscarrollbrasil.com.br

And now appears a mystic word, but if it be inverted,

We find the ending of this book in plainest text asserted.

This, the earliest known ambigram, is from Peter Newell's book *Topsys & Turvys* (1893). His 1901 illustrations to Carroll's books were used by Martin Gardner in *More Annotated Alice*.